"The aim of this book is to give a general account of the fungi and their impact upon us—how they live and grow and reproduce, what they do and how they do it, what their place is in the general scheme of nature. It is written in a form understandable by the layman and beginner, and is intended mainly for those who wish to learn about the how and why and when and where of fungi. The readers for whom it is designed include not only beginning students in mycology, whether amateur or professional, but all those who, either for practical reasons or for pure enjoyment, might want to get acquainted with these prevalent and interesting and often important plants."
From the Prefatory Note

This third revised edition restores the chapter on classification which had been left out of the second edition, and in addition contains a new chapter on the toxicity of fungi.

Clyde M. Christensen is Professor of Plant Pathology at the University of Minnesota.

T0136095

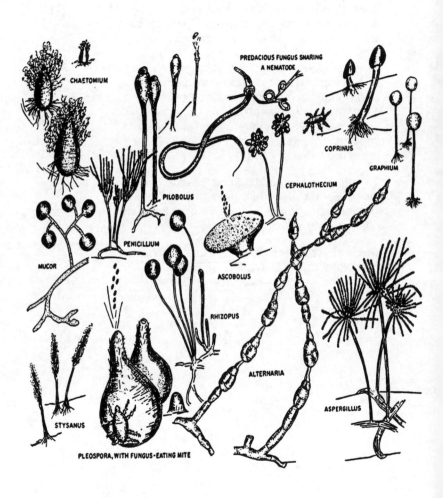

CHAETOMIUM

PREDACIOUS FUNGUS SNARING
A NEMATODE

COPRINUS

GRAPHIUM

PILOBOLUS

CEPHALOTHECIUM

PENICILLIUM

MUCOR

ASCOBOLUS

RHIZOPUS

ALTERNARIA

ASPERGILLUS

STYSANUS

PLEOSPORA, WITH FUNGUS-EATING MITE

University of
Minnesota Press

The Molds and Man
An Introduction to the Fungi
Clyde M. Christensen

McGraw-Hill Book Company
New York   Sydney   Toronto   London

THE aim of this book is to give a general account of the fungi and their impact upon us — how they live and grow and reproduce, what they do and how they do it, what their place is in the general scheme of nature. It is written in a form understandable by the layman and beginner, and is intended mainly for those who wish to learn about the how and why and when and where of fungi. The readers for whom it is designed include not only beginning students in mycology, whether amateur or professional, but all those who, either for practical reasons or for pure enjoyment, might want to get acquainted with these prevalent and interesting and often important plants.

A few of the larger fungi have been known from casual observation for thousands of years, but as living, growing plants they have been studied for only a bit more than two centuries. During most of this time mycology — the study of fungi — has been for mycologists only. No one else was much interested in how fungi grew, or where or why.

In the last few decades this has changed. The mycological horizons have begun to expand rapidly and in some unusual and unexpected directions. Fungi have been found to be of basic and practical significance in various fields, from human medicine to grain storage, from agriculture to architecture, from fundamental studies on the nature of sex to biological warfare. There are few phases of our life, apparently, into which fungi do not enter, for good or ill. This new importance does not mean that fungi have suddenly come into their own, for they've always been there. It is just that we finally have begun to learn more about the details of the various and often subtle ways in which they influence our lives.

One result of this increasing awareness of the part the fungi play in the world at large and in our everyday practical economy is that many who have little or no mycological background want to gain at least a speaking and working knowledge of them. To get it has been rather difficult. General accounts of the fungi are given in many botany texts, but these are necessarily brief summaries of the main groups; and the mycology texts now available have been written principally for advanced, professional students. It is hoped that this book will fill the gap between the two, and will introduce the fungi not only to those who expect to go on to more advanced work in mycology but also to those who just want sufficient information about the fungi to be able to talk and think intelligently about them.

Throughout the book I have tried to include some of the arresting, unusual, little known, and often wonderful aspects of the common fungi. To me, the study of fungi has been more than just a professional occupation — it has been absorbing, enriching, stimulating adventure. I should be glad if some of this feeling, at least, I had managed to pass on to the reader.

C. M. C.

TABLE OF CONTENTS

List of Plates

The Molds and Man

CHAPTER I

What Fungi Do and How They Grow

THERE are about two million different kinds of living things on the earth, of which fungi make up approximately from eighty thousand to a hundred thousand species. Many of these fungi are so prevalent and abundant that they must be considered one of the more successful forms of life. If most of us are not aware of the multitudes of fungi about us all the time, it is only because we have neither the background nor the technics necessary to see them. They are small and inconspicuous, but they make up for this in other ways that allow them to roll with the blows of environment or fate more adeptly than man or many other higher animals are able to do. If they have not yet exactly inherited the earth, they do eventually inherit a goodly portion of the living things on the earth, as well as the goods and products made from these living things. They affect us and other living things in many different ways. Some of these are discussed in the following chapters, but a brief summary of their major fields of activity will serve to introduce the reader to the general activities of fungi.

What Fungi Do

Fungi are among the chief causes of disease in plants. Cultivated plants always have been, and doubtless will long continue to be, one of our basic sources of wealth and well-being. No stable economy is possible without the control of plant diseases.

These diseases are not controlled by governmental edicts, nor do they hearken much to the noises made by legislators. Basic and applied research in plant pathology has made a start in the control of some of the most destructive plant diseases: we have made a couple of solid hits, as well as a few fouls; but we are hardly past the second inning in the control of fungus diseases of plants.

Fungi also cause heavy loss in stored seeds of all kinds, especially when they are stored or shipped in bulk. It is commonly believed that once seed is harvested it is safe from loss. This is not true. Grain-infesting insects and storage fungi take a heavy toll of stored grains. The storage fungi are especially insidious because they invade grains stored at moisture contents that practical grain men consider safe and often cause serious damage before their presence is even suspected.

Manufactured products of many kinds are very subject to attack by fungi. Fungi decay wood, cloth, fabrics, twine, electrical insulation, leather, food products of all sorts, and a variety of other materials. Molds even corrode the lenses of binoculars, telescopes, and microscopes if the air is sufficiently humid to permit them to grow rapidly. About the only substances not attacked, broken down, and decayed by fungi are metals.

Fungi also are responsible for a great variety of diseases of animals, and it is only in recent years that their importance in the field of human medicine has come to be recognized. An understanding of the nature of fungi may not be so vital to the student of human medicine or veterinary medicine as it is to the plant pathologist, but some background in the fungi is likely to be helpful, since it aids him to detect fungus infections when he encounters them.

Not all fungi are harmful, and a few of them have been exploited for our benefit in one way or another. Some are grown commercially for the production of drugs, organic acids, enzymes, and feed supplements. They are used in the production of certain kinds of cheeses, of soy sauce, and of several other foods.

The above is sufficient to indicate that fungi do affect us directly, that they are of some concern to all of us. Indirectly, they are probably of even greater significance, for they play an important part in the general scheme of nature. Most fungi are scavengers. They grow upon and decay the remains of plants and animals, and convert these into rich soil. It is a common scientific truism that without green plants there would be no life on the earth, since green plants are almost the only living things able to use the energy from the sun to manufacture products on which all other life depends. It is equally true that without fungi most green plants could hardly survive, since they depend upon the products of fungus decay in the soil.

Besides their direct and indirect effects upon us and upon other living things of importance to us, fungi may be interesting just in themselves. Some of them have formed close partnerships with other plants, from algae to orchids and forest trees. A few have formed partnerships with insects, both members sacrificing independence for security. There are still others that are of interest to any student of biology because of their wonderfully subtle, complex, and accurate adaptations to the life they lead. Fungi may be simple in structure, but most of them are far more complex than appears on first glance. We should not be too much impressed by the fact that we have placed ourselves near the top of the evolutionary ladder and the molds near the bottom. As a group the fungi are eminently successful; they have what it takes to survive. Molds were on the scene and doing well long before man appeared, and many of them are pretty certain to be around in the remote future, long after man has played his part and groped his way into the dark wings or fallen through the trapdoor on the stage.

How Fungi Grow

To understand how fungi have got where they have in the world and how they are able to influence us and other forms of life in the ways they do, it is necessary to know something about how they grow and reproduce and get around, what problems they encounter, how they surmount these problems, and what happens to them when they get in a jam. The present chapter summarizes the lives and habits of fungi or molds in general. It must inevitably leave much territory unexplored and unexplained, but once a person has absorbed the facts and implications in this summary, he should have a fairly good idea of what fungi are and how they operate. Some of the gaps will be filled in with detailed accounts of individual fungi later in the book.

The fungi as a group share the following three characteristics:

1. They have no chlorophyll. This means that they cannot manufacture their own organic food, such as sugars, starches, celluloses, proteins, and fats, as the green plants do. Therefore they must live on the remains of other plants or animals or on living plants or animals; in other words, they must live either as saprophytes or parasites.

2. The growing, food-getting part of a fungus is made up of long, hollow, branched cells which in aggregate are called mycelium. Mycelium will be described and illustrated in some detail shortly.

3. Fungi reproduce by means of spores. Some of the higher plants, such as ferns, also reproduce by spores of a sort, and certain bacteria form spores to tide them over unfavorable conditions; but the fungi have gone into spore production in a really big way as a means of increasing their kind. This is one of the reasons for their success. Spore production as developed by many of the fungi makes for rapid increase, far travel, and wide distribution, and almost no other forms of life can equal them in these capacities. Spores will be taken up in more detail in Chapter 2.

These three characteristics — (1) no chlorophyll, (2) mycelium, and (3) spores — are the only ones that nearly all fungi have in common. Not all of them have all three of these. Yeasts are fungi, but do not form mycelium; they grow by a process called budding. A number of common molds, when growing in liquids, do not form mycelium either, but grow by budding as the yeasts do. There are also some half dozen or so fungi that are not known to form spores. One of the satisfying things about biology is that there are exceptions to any general rule that the ingenuity of man can devise. However, with the exceptions noted, all molds or fungi are known by the three simple characteristics just given. All of these characteristics will be enlarged upon in the following pages, because they determine to a large extent the how, where, when, and why of fungus life.

LACK OF CHLOROPHYLL

Fungi are not the only plants that have no chlorophyll. Bacteria, slime molds, and a few other kinds of lower plants share this character with them. We are inclined to look down upon all such plants as something on the other side of the biological tracks, low forms of life that depend upon the bounty of their more self-sufficient and industrious relatives. Man enjoys arranging living things into a sort of social hierarchy, with himself always at the top. It is quite true that molds, as well as men, are dependent upon green plants. The fungi have at least the virtue of contributing as much to the support of the green plants as the green plants do to them, which is more than can be said of *Homo sapiens*. Without the rot and decay of the scavenging fungi, even the green plants would soon go by the board. As factors that make the biological world go round, the fungi and the green plants are about equally indispensable.

Most fungi are saprophytes. *Saprophyte* means, literally, *rotten plant.* By derivation it signifies a plant that lives on dead material, which it consumes and uses for food. Such consumption of plant or animal material by saprophytic fungi we often designate as rot or decay. We rot or decay the food we eat, too. The majority of fungi are *obligate* saprophytes — that is, they are obligated to live that way, they can live only on dead material. So long as the fungi are living on and converting the remains of plants and animals into fertile soil, they are performing a beneficial and necessary service for the biological community. This is their main sphere of action, and it probably will continue to be their main sphere of action until life disappears from the earth.

Not all saprophytic fungi are beneficial, however. Those which

7

grow in our manufactured and processed goods and convert them into soil often are costly pests. For example, the fungi that normally grow in fallen logs in the forest and convert the wood slowly into rich humus grow with equal facility in the sills, studdings, and rafters of privies and palaces, the hulls and planking of wooden ships, and the staves of beer barrels. To them wood is wood, food to be converted into more fungus, and they rot the houses of the well-to-do as well as the hovels of the poor.

The fungi that normally grow in dead stems and leaves of grass grow with equal avidity in paper and clothing, shoes and shower curtains, electrical insulation, girdles, bread, and glue. It is this saprophytic mode of life, the ability to consume plant and animal products of any and all kinds, that makes some of the common fungi as prevalent as they are and such a problem in so many different products and places.

Some fungi are parasites. *Parasite* means, literally, *beside the food*. Technically, a parasite is an organism that lives on or in another plant or animal and gets its food from the plant or animal on or in which it lives. Many diseases of human beings are caused by parasites that live on or within our bodies and get their nourishment from us. Similarly many diseases of plants and animals are caused by mold parasites that live on or in them.

Several groups of fungi are *obligate* parasites, notably the downy mildews, the powdery mildews, and the rusts. They are obligate parasites because they can live only in the living tissue of the particular host to which they have become adapted; they are obligated to live that way. Some of the obligate parasites among the fungi have evolved a rather extreme specialization as to the host on which they will grow — that is, a given species of obligately parasitic fungus, such as the one causing stem rust of wheat, is not just a uniform species. One race of this fungus will grow on certain varieties of wheat, others will grow on other varieties of wheat. Each is able to attack only particular and special hosts. This complication will be gone into in more detail when plant diseases are considered in Chapter 6. It is perhaps worth knowing that the study of this specialization in some of the obligate parasites among the fungi has proved to be basic not only to the control of plant diseases but also to an understanding of some of the diseases of man caused by parasites, and has implications that extend to many other fields also. The specialization is one example of the almost infinite variation that occurs in nature, and, in passing, it allows one to see evolution in progress.

Many of the fungi can live either as parasites or as saprophytes, regulating their life according to the opportunities offered. These are called *facultative* parasites or *facultative* saprophytes. The majority of fungi that cause diseases of plants and animals are of this nature. They can grow on plant refuse or in living plants. Up to recent times it was believed that certain crops quickly "exhausted" the soil and that therefore they could be grown on the same ground only for a few successive years. The so-called soil exhaustion that results from growing flax, bananas, cotton, alfalfa, and many other crops year after year on the same soil is not soil exhaustion at all. It is merely the result of the accumulation of facultatively parasitic fungi in the soil, to the point where the fungi eventually eliminate the plant. Some of these facultative parasites among the fungi are also highly specialized as to the host they will attack, and our knowledge of this is now being put to profitable use in agriculture and in other fields.

In summary, molds can live as obligate saprophytes, in which case they are restricted to dead materials. This includes not only dead plant and animal material as such, but most of our goods and chattels. They can live as obligate parasites, on living plants or animals solely, and then are likely to be very highly specialized as to the varieties of hosts on which they grow. And they can live as facultative parasites or facultative saprophytes, getting their food from either dead or living things. Practically all the living plants and animals in the world and all the products of these serve as food for fungi. The fungi live on, decay, and consume almost everything. In doing so, they must break down complex products into simple ones, absorb these, and then convert the absorbed material into more fungus. This is where mycelium comes in.

MYCELIUM

What mycelium is. We can best understand what mycelium is if we look at it with the naked eye and with a microscope and watch it patiently for hours or days to see how it develops and grows. As a poor substitute for the actual observation of growing mycelium, the following account of that growth is offered.

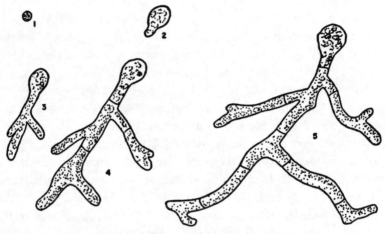

Figure 1. The development and growth of mycelium. (1) A dormant spore. (2) A germinating spore. The spore has increased in size, and a protuberance, the germ tube, is growing out of one side. (3) The germ tube has increased in length, and two short branches have grown from it. (4) Crosswalls have begun to form. Several branches have grown out, each one known as a hypha. (5) The mycelium is now on its way, and will continue to grow by terminal elongation of the cells.

Suppose we put a bit of leaf mold or soil in a dish of water or on an agar or gelatin medium. Within from twelve to twenty-four hours mycelium will begin to grow out. This mycelium comes from mycelium or spores present in the leaf mold or soil. The mycelium grows out in the form of long cells, as illustrated in Figure 1. Each branch puts out new branches at intervals, and these branch again. The only growth is at the ends of the cells. Mycelium advances by terminal growth. The mycelium of a fast-growing fungus will grow across the field of the microscope as we watch it; we can actually see it advance. This advance involves not a mere extension or stretching of the cell, but a continual formation of new cell material, new cell walls. We do not know how cell walls are formed as rapidly as they are, but we definitely can see that they are formed rapidly.

Given plenty of food and no competition from other fungi or bacteria, a fast-growing fungus will cover the surface of a three-inch-wide culture dish in two or three days. The weight of this mycelium may be inconsequential, but the length of cells formed is somewhat surprising. A moderately fast growing fungus will advance at the rate of about 1/8,000 inch a minute. Each advancing cell will put out a new side branch on the average of from every thirty to forty minutes, and each one of these will advance at the same rate and produce new branches. In twenty-four hours a fungus colony such as this can produce a total length of over one-half mile of mycelium, and in forty-eight hours will form a total length of hundreds of miles of cells. This rapid growth of mycelium explains why molds can invade materials rather rapidly when conditions for their growth are favorable. They can grow through a loaf of bread in a few days. They can grow rapidly in cream, and if butter is made from cream that has not been kept clean and cool, it may contain several miles of mold mycelium per pound.

Different fungi produce mycelium of different sorts and sizes and in different amounts. Some of the water molds produce only a tiny tuft of sparsely branched mycelium that is scarcely visible even with a microscope. Some of the water molds have mycelium with cells so large that they can be seen with the naked eye. The mycelium of mushrooms and of wood-rotting fungi extends for feet or yards through soil or decaying wood. Through these miniature pipelines the essentials of life and growth are conducted at a speed that is amazing when one considers the friction that must be overcome in the flow of relatively viscous protoplasm through long, narrow cells of such a small diameter and large wall surface. (See Plate 1 facing page 88.)

In one relatively large group of fungi the mycelium, regardless of how extensive it is, has few or no crosswalls, and so there may be literally miles of mycelium consisting of a single cell. Within this single, much branched cell, there may be considerable specialization, the terminal portions concerned mostly with the getting of more and more food, the older portions mostly with spore production. In some ways such a cell is simple, but in other ways it is too complex for us to have learned much about how it functions.

The structure and function of mycelium. Each cell of the mycelium consists of a wall which encloses protoplasm and one or more nuclei. The wall is composed largely of cellulose, similar to the cellulose in the cell walls of most other kinds of plants. The diameter of the cell will vary with the fungus and with the conditions under which it is growing. The largest are from 10 to 20 microns in diameter. A micron is 1/1,000 millimeter, or 1/25,000 inch, so that such a cell would be from 1/2,500 to 1/1,250 inch in diameter. The smallest are less than 1/2 micron, or 1/50,000 inch in diameter. We occasionally encounter mycelium that does not exceed 1/100,000 inch in diameter. Within this narrow space are included a cell wall and its enclosed protoplasm. The materials of life are apparently transported with relative ease through these exceedingly small conduits.

The individual cells of mycelium, like the individual cells of almost any living thing, are minute chemical factories in which the mysteries of life go on at a great rate. The protoplasm which fills the cells at the growing tips of the mycelium manufactures dozens of different kinds of enzymes and a number of simple and not-so-simple organic acids. These enzymes and acids diffuse out through the thin wall into the material through which the fungus is growing. If the fungus is growing in wood, these digestive chemicals diffuse out into the wood, break it down into simple sugars, and these diffuse into the mycelium, where they are converted into more fungus. If the fungus is growing in bread, the enzymes and acids diffuse out into the bread, break down the celluloses, starches, sugars, proteins, fats, and other constituents of the bread into simple compounds, which diffuse into the fungus cells and there furnish food and energy for more growth.

The essential difference between our digestive system and that of fungi is that the fungi digest most of their food outside of their bodies. The basic processes of digestion are the same in molds as in men. Both use enzymes and acids to convert complex materials into simple ones. A fungus enzyme, taka-diastase, is, in fact, sometimes taken by human beings whose digestive processes are not what they should be. The main difference between the digestive processes of men and of molds is that the fungi digest their bread before they eat it, while we eat it first, then try to digest it. The actual process of digestion is essentially the same in both.

Our digestive system involves such complex things as teeth, esophagus, stomach, intestines, plus many associated organs and processes. It is over-complex. Things go wrong with it from one end to the other. Fungi are free of these evils. They get no tooth aches or stomach aches, they are never observably faint or dull from overeating, they are not subject to the miseries and embarrassments of indigestion, stopped-up bowels, or flatulence. If we could stick our fingers into our soup or bread or meat and absorb it forthwith, we should be eating as the molds do. Their system may be a "simple" one, but a hundred thousand different kinds of fungi have been doing all right with it for a very long time, so it can hardly be considered unsuccessful.

Common, ordinary, everyday fungi secrete a great number of different digestive enzymes. This enables them to live on and digest a wide variety of materials. Some animals are called omnivorous because they can subsist on both animal and vegetable food, but they are fussy feeders compared with some of the molds. Penicillium, for example, can subsist on the remains of thousands of different kinds of plants, on cloth, leather, paper, wood, tree bark, cork, animal dung, animal and insect carcasses, ink, syrup, seeds of all kinds, manufactured cereal products and the boxes in which they are packed, including the wax and ink on the outside, on stored fruits and vegetables, soil, glue, paint, liquid drugs, hair and wool of all kinds, on the wax in our ears, and on literally thousands of other common products. Not all of these things furnish it with a well-rounded diet, to be sure, but it can live, grow, and reproduce on and in them with sufficient regularity so that we encounter it on them commonly. Penicillium is an eminently successful organism, able to make its own way in the world with no favors or handouts. In the present age that alone should warrant some respect.

THE INFLUENCE OF ENVIRONMENT UPON GROWTH

Fungi are influenced by such things as temperature, water, oxygen, acidity-alkalinity, food, minerals, vitamins, growth-promoting substances, and various subtle etceteras. Like other small and structurally simple things, they are much more children of their environment than we are. Some of them have conspicuous weaknesses in this respect, which is fortunate, because it enables us to control or eliminate them by making it too hot for them or too cold, too wet or too dry. They are likely to be somewhat more adaptable, however, than most of us realize. Some of the major factors that influence the growth of common fungi will be discussed briefly.

Temperature. Fungi have no thermostatic controls, such as we and most other higher animals have, to regulate their temperature. Yet they can adjust themselves to changes more readily than the more complex forms of life.

In general, most fungi grow best between 70 and 90 degrees Fahrenheit. At 50 degrees these fungi will grow somewhat slower than they do at from 70 to 90 degrees, and when it gets down to 30 or 40 degrees they stop growing. They do not die. They simply become dormant and wait for the better times that are sure to be just around the corner. Most of the ordinary fungi will easily survive freezing for months or years. Many are not injured by the temperature of liquid air, and some have been exposed to a temperature close to absolute zero without loss of viability. Low temperature is not effective in eliminating most fungi. They stop growing when it becomes cold, but when the temperature rises again, they resume growth almost at once.

Many fungi can grow — not only live, but actually grow and develop and produce mycelium and spores — at temperatures below freezing. Meat and other foods stored in freezers must be kept below 20 degrees Fahrenheit — well below freezing, that is — to prevent their becoming moldy. Foods stored in ordinary home refrigerators will become moldy rather rapidly. A side of beef stored in a butcher's locker may, within a few weeks, become covered with a downy fungus fuzz. Such mold presumably contributes to the tenderness and flavor of the meat — if it is the right kind of mold. In the transport of refrigerated meat from Australia to England it is not unusual for fungi to develop sufficiently in the meat to cause some reduction in quality, mainly by spotty discoloration.

Fungi also grow in and rot flower bulbs in frozen ground, kill perennial forage grasses and legumes during the winter, and ruin golf greens when they are covered with snow. A few of these fungi are so adapted to a life at low temperature that they do not grow well much above freezing: they are active in the northern regions only in late fall and early spring, when almost all other things are dormant. Plants whose roots are solidly frozen apparently have little protection against invasion by these low-temperature fungi.

High temperature, however, is another story. A fungus which grows well at 80 degrees Fahrenheit will begin to slow down at from 100 to 110, and if it is exposed to 130 to 150 degrees for hours or days, it will reluctantly give up the ghost. A temperature approaching that of boiling water (212 degrees Fahrenheit) finishes off most fungi instantaneously. Few of them can endure 160 degrees for more than seconds or minutes.

This sensitiveness of most fungi to high temperature is one of their weaknesses, and it is taken advantage of in agriculture and industry. Some parasitic fungi are eliminated from the seeds of crop plants by heat treatment. Foods processed by heat — boiling, baking, or pressure cooking — are free of living fungi. They may, however, become contaminated later. As an example, even the highest grades of wheat flour contain a moderate number of living spores of different kinds of fungi; these are killed when bread is baked, but bread and other bakery goods may become heavily contaminated with molds soon after they are out of the oven.

Some fungi like a moderately high temperature, and begin to hit their stride only between 120 and 130 degrees Fahrenheit — so hot that most plants and animals, including man, could not survive in it for more than a few hours at most. These high-temperature molds are not very numerous in the way of species, but they are exceedingly common and widespread. A number of them are involved in the preliminary heating that occurs when grain, hay, cotton, and other materials are stored moist in large bulks. Under the right conditions the preliminary heating due to fungi may be followed by thermophilic (heat-loving) bacteria, which can raise the temperature up to about 170 degrees Fahrenheit, and this rise may be followed by purely chemical heating to the point of combustion. This will be gone into in more detail in Chapter 7.

Water. Some of the fungi grow only when immersed in water. They are common and apparently nearly ever-present in streams, lakes, ponds, and moist soil. There they live on a great variety of living and dead plants and animals, and probably make up a rather significant part of the flora. A few of them are parasitic on fish and other aquatic animals. Relatives of these water molds are found in nearly all kinds of soil, and some are destructive parasites of the roots of many wild and cultivated plants. They are especially difficult to contend with, since they often can parasitize a wide range of host plants, both wild and cultivated, and can also exist as saprophytes on debris in the soil.

At the other extreme, some of the most common and widely distributed of fungi grow vigorously in seeds, flour, wood, leather, and almost innumerable other products that contain only from 12 to 15 per cent water. The lower limit of growth of these drought-resistant fungi is a water content in equilibrium with a relative humidity in the surrounding air of from 70 to 75 per cent. This means that whenever the relative humidity of the air exceeds 70 per cent, these particular fungi can begin to grow and cause deterioration; usually the result is trouble of one sort or another.

Between the tropics of Capricorn and Cancer the relative humidity of the air often exceeds 70 per cent throughout much of the year. Frequently it is above 75 or 80 per cent for days or weeks at a time. In humid, warm regions, such as the Gulf Coast of the United States, these fungi constitute a major, if often unrecognized, factor in everyday life, in that they rot at a terrific rate shoes and clothing, equipment, food, and stored goods of all kinds. Some economic planners have suggested that our industries and agriculture eventually will have to move to the tropics, when we have exhausted most of our stored energy in the way of coal and oil. Fungi would not make such a move impossible, but they certainly would add greatly to the difficulty and cost of maintaining what we have come to consider a normal life, not to mention what they would add to our discomfort.

Warm, humid conditions are not limited to the tropics. A research worker for one of the larger manufacturers of electrical equipment stated, during the Second World War, that much of the mold deterioration of the goods they shipped to the armed forces occurred before the shipping crates ever left the state of New York. Similar warm, humid, and essentially tropical conditions are common in basements in the northern countries during moist weather, in the holds of ships, in clothes hampers, breadboxes, within the enclosed milling machinery where grain is reduced to flour, in the rooms where macaroni and spaghetti are dried, in spinning and paper mills, within walls where condensation occurs or rainwater has seeped in, in your shoes, in clothing next to perspiring skin, and in many other places. The fungi take advantage of this fact.

The essential point is that whenever and wherever the relative humidity exceeds 70 per cent regularly, slow molding will occur, and when it exceeds 75 per cent, molding will occur within a matter of a week or two in practically all the materials that are subject to fungus deterioration, which is well-nigh everything but metals. Sometimes this molding is not more than a minor annoyance. Sometimes it is a major factor in the economics of raw and manufactured foods and feeds, or in the length and quality of service to be expected from fabrics, fibers, wood products, electrical equipment, optical apparatus, and so on.

Oxygen. Nearly all fungi require oxygen in order to live and grow, although there are exceptions to this. Some of the yeasts are anaerobic; that is, they do not need oxygen from the air. A few of the other fungi can do with a very small amount of oxygen and can tolerate a high concentration of carbon dioxide, but they are not many.

Most of the fungi not only need oxygen, even if only in small amounts, but they are poisoned by carbon dioxide, even as we are. Very few of them can grow if the concentration of carbon dioxide in the air exceeds 50 per cent, regardless of how much oxygen is present. When they are growing in soil, stored grain, baled cotton, and other bulk materials, they may be inhibited or killed by the carbon dioxide they themselves produce. They are tougher than we are in this respect, since they can live for days or weeks in a concentration of carbon dioxide that would kill us in seconds or minutes; but to them also carbon dioxide in too great a concentration definitely is poison. There are no obligate anaerobes among the fungi, as there are among the bacteria, some of which will grow only in the entire absence of free oxygen.

Food. Like all other living things, molds need food. It should be obvious to anyone who has read this far that most of the common materials around us furnish an ample diet for many different kinds of fungi. Some fungi thrive on an extremely wide range of food, and seem capable of digesting and decaying almost everything but metals. Others are so specialized that they will grow only on specific kinds of pollen grains, on hair or feathers, in the wood of certain kinds of trees, on the leaves of a single variety of wheat or rice, on the left hind leg of a particular species of water beetle, within the body of certain kinds of insects, on the dung of certain kinds of animals, on the fine, absorbing roots of certain kinds of orchids or forest trees, or in other peculiar and special environments.

What it amounts to is that the fungi are experimenting with many different ways of getting food. Some of them have become such extreme specialists that one wonders how they manage to survive; others consume so wide a range of materials that one wonders how the world has escaped as long as it has from being converted into a gigantic mold colony. As it is, the fungi have done fairly well at converting a major portion of the world into mold. Given a slight but consistent increase in temperature and relative humidity over a large portion of the globe for a few eons, probably they would become a dominant form of plant life.

Light. Light is only a minor factor in the growth of most fungi. A few of them need a certain amount of a certain quality of light in order to reproduce normally, but the majority of them grow as well in darkness as in light. Ultraviolet light will stimulate some, inhibit others, or kill them, its effect depending on such things as wave length, time of exposure, and the particular fungus involved. The majority of fungus spores are more resistant to injury from ultraviolet rays than are bacteria. Some fungi have a dark pigment in their mycelium and spores that apparently protects them from injury by ultraviolet light.

Poisons. Since plants and processed goods are often protected from fungus attack by being covered or impregnated with chemicals toxic to fungi, and since the lay public often enquires about fungicides for different purposes, a brief discussion of this aspect of the subject may be in order. The field is so large and complex, and so many qualifications attach to nearly any general statement that can be made, that the writer takes it up at all only with considerable trepidation. The following, however, seems reasonably conservative.

First, there is no one general, all-purpose fungicide. Literally thousands of different compounds have been manufactured and tested as fungicides, and a number of these are very effective for certain purposes. Creosote has been used as a wood preservative for more than a hundred years. There is no question as to its effectiveness in preserving wood from decay. Yet one fungus is known that grows almost only in wood impregnated with coal tar creosote. This fungus does not harm the wood, but it will grow on and in creosoted wood, and even on creosote alone: it can use the creosote for food. Inorganic salts of copper are excellent fungicides for some purposes, and are widely used, but again there are fungi that can grow in relatively high concentrations of copper salts. Sulfur is very toxic to rust fungi, but to many others not particularly so.

A number of rather complex chemicals have been devised for what are called "seed treatments." They are applied to agricultural and garden seeds to prevent fungi from attacking the germinating seed and young seedling. Some of these are effective on one kind of seed, almost useless on another kind of seed. Or they will prevent seedling decay on one crop in one region, but not in another region in the same year, or prevent it well on one crop in one region in a given year, but be relatively ineffective the next year. Some of these seed treatment compounds, known to be highly toxic to fungi when sufficient water is available to dissolve them, are totally ineffective in preventing the molding of stored seed where no free water is present. They can be applied heavily enough to cover the seed with a coating of fungicide, yet heads of mold will grow out with visible bits of these very toxic fungicides on them.

One fungicide may be excellent for fabrics, another for leather, another for stored paper pulp. The effectiveness of any one of these compounds in controlling a given fungus, under a specific set of conditions, can be determined only by testing it under those conditions. Laboratory tests in which the fungicide is placed in a liquid or agar medium are likely to be relatively worthless in evaluating its fungicidal properties in practice.

It is a common misconception that ordinary oil paints, and especially asphalt paints, are fungicidal. People cover tree wounds, or the exposed stubs left when branches are removed, with these paints, in the fond hope that they will prevent fungi from entering and decaying the tree. Ordinary paints are not fungicidal:

fungi grow in, on, and through most of them with ease — actually use such paint for food. Under certain conditions, coating wood with paint will stimulate decay rather than prevent it. As will be seen in Chapter 6, ordinary asphalt paint, far from being a tree protective, has in the past, upon at least one notorious occasion, been responsible for inoculating healthy trees with fungus disease.

There is, then, no one general fungicide. What is poison for one fungus may be food for another. A compound that serves as an effective fungicide under one set of conditions may be useless under another: the effectiveness is influenced by a large number of factors. Rapid strides have been made in the last two decades in the development of new fungicides for specific purposes. Effective fungicides are now used in cold creams, bread, butter wrappers, cheese, fabrics, leather, paper, wood, agricultural seeds, and various other things subject to attack by fungi. But before the layman starts out on his own to use any of these compounds, he would do well to get advice from someone who knows his way around with the specific problem at hand.

Summary

There are close to a hundred thousand different kinds of fungi. They are not rare, chancy plants growing just here and there, but are present everywhere throughout the world, growing on all conceivable kinds of things. Their size is small, their structure simple, but they have many of the characteristics that make for survival, and on the basis of numbers and prevalence they must be counted among the successful and dominant organisms on the earth.

They affect us in many different ways. Directly, in that they cause diseases in our plants and in animals, including man. They affect us indirectly in that they are one of the major groups of organisms responsible for converting the dead plants and animals of today into the rich soil of tomorrow. In life as we encounter it on the earth, the fungi are probably about as important as the green plants.

Fungi have no chlorophyll, and therefore can not use the energy of the sun directly, as the green plants do. They must live either as saprophytes, on dead material, or as parasites, on living material. Many of them can live either as parasites or as saprophytes, as the occasion offers.

The growth of fungi is affected by temperature, water, oxygen, food, toxic materials, and some other factors not discussed above. Among the common fungi are those that endure less water, a lower temperature, a higher temperature, or a higher concentration of toxic materials than almost any other organisms on earth. Some of them grow under a very limited range of conditions, others are encountered almost everywhere.

The summary of the summary is that one can best learn about fungi not from a general discussion of them, but from a study of how the individual fungi live and grow and reproduce. Which brings us to the next chapter.

Additional Reading

M. J. Berkeley, *Outlines of British Fungology*. London: Lovell Reeve, 1860. 442 pp.

E. A. Bessey, *Morphology and Taxonomy of Fungi*. Philadelphia: Blakiston Company, 1950. 791 pp.

M. C. Cooke, *Fungi: Their Nature, Influence, and Uses*. London: Kegan Paul, Trench, Trübner & Co., 1894. 299 pp.

E. C. Large, *The Advance of the Fungi*. New York: Henry Holt & Company, 1940. 488 pp.

R. T. Rolfe and F. W. Rolfe, *The Romance of the Fungus World: An Account of Fungus Life in Its Numerous Guises, Both Real and Legendary*. London: Chapman & Hall, 1925. 309 pp.

CHAPTER 2

The Reproduction and Dissemination of Fungi

LIKE other living things, molds or fungi reproduce their kind, and the ways in which they go about making more molds help to determine when, where, how, and on what or whom they grow. All but a very scant few of the fungi reproduce by means of spores. How spores are formed and how they get around are described in the present chapter. From the practical standpoint, it will be seen, there is a good deal of significance in these methods of increase and dissemination.

How Spores Are Formed

The best way to find out how fungi form spores is to watch the process under a microscope. As a substitute for this, I shall describe and illustrate a few typical cases.

Alternaria is the name of a very common fungus that grows on living plants, straw, leaves, grass clippings, rotten fruit, seeds, and similar plant remains, all of which it changes into Alternaria spores with single-minded speed and efficiency. The mycelium of Alternaria, after growing for a few days and storing up a modicum of food, sends up a tangle of aerial branches that divide into cells. These soon grow into spores of different shapes and sizes (see Plate 2). The typical spores are tadpole shaped, several celled, with crosswalls running in both directions. But the fungus may form single-celled and two-celled spores. All of these are illustrated in Figure 2. If we catch dark-colored, tadpole-shaped spores with crosswalls on a slide exposed to the air, we can be certain that they are spores of Alternaria. Since this fungus is a common cause of respiratory allergy in human beings, such recognition is of more than academic importance.

Alternaria forms its spores in prodigious numbers just by division of aerial mycelium into cells and the later growth of these cells into spores. These spores are so small and buoyant that infi-

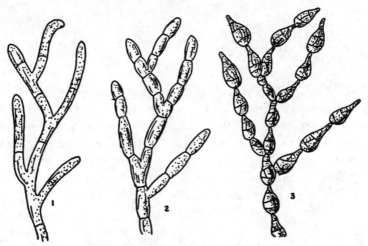

Figure 2. Spore production by Alternaria. (1) Branched hyphae grow up into the air. (2) Crosswalls form at short intervals. (3) Tadpole-shaped spores develop from these cells.

nitely small currents of air can pick them up and whisk them away. They are often so abundant that invisible clouds of them, said to amount to tons of spores, are encountered in the air, drifting across the country as discrete but invisible clouds. Our air, indoors or out, seldom is free from spores of Alternaria.

Penicillium means *a diminutive brush* or *broom,* descriptive of the cluster of spore-producing branches. Within a few days after a new colony of Penicillium has begun to grow, on any of the infinite number of things Penicillium can grow on, special stalks grow up from the mycelium, the tips of these stalks branch, and branch again, and finally each branch produces a long row of round or oval spores. The process is illustrated in Figure 3. A single brush of spores is just large enough to be visible to the naked eye, but the individual spores are only from 2 to 5 microns — from 1/8,000 to 1/5,000 inch — in diameter. Each spore consists of a single transparent cell. Masses of spores of Penicillium are likely to be some shade of green, although other colors occur.

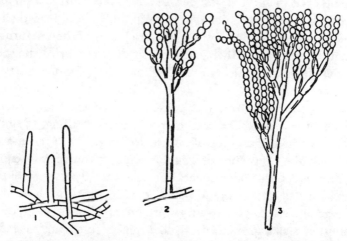

Figure 3. Spore production by Penicillium. (1) Single stalks grow vertically into the air. (2) A single spore-producing stalk has formed a characteristic "brush" at the tip, and chains of spores are being formed on the end of each branch. (3) The brush increases in size, and each branch gives rise to a long chain of spores.

A few of these spores falling on a favorable spot will germinate within hours, produce a visible patch of mycelium within several days, and in less than a week will have resulted in hundreds of millions of spores, each of which can repeat the process. We commonly think that some of our insect pests, such as flies and mosquitoes and cockroaches, are fairly prolific, but compared with Penicillium and many other fungi, these pests are slow, stodgy, and conservative. On a moldy orange in the grocer's window (but seldom in the top layer) the green spores of Penicillium will be so thick that you can rub them off as a heavy powder. The slightest disturbance of the orange or of the air around it will send up a cloud of millions of spores. If you merely touch such a fruit, you carry home hundreds of thousands of spores on your hand and transmit them to everything you handle as well as to many things you don't. This does not matter very much, because the more or less invisible bits of moldy stuff around even the neatest and cleanest of houses furnish a sufficient supply of spores of Penicillium to keep the air well loaded all the time. The very visible moldy material out of doors furnishes countless numbers of them, and it is small wonder that the spores of this fungus fill the air throughout the world, winter and summer.

Rhizopus means *root foot*, descriptive of the short, rootlike strands of mycelium that anchor the elegant, swaying, top-heavy stalks on which the heads of spores are borne. On agar in a culture dish the mycelium of Rhizopus will grow rapidly enough so that under the microscope one can actually see it advance. Starting from one or a few spores, it will form a patch of mycelium an inch across in twenty-four hours. From this mycelium clumps of stalks grow upward, like miniature clumps of cottonwood or basswood trees around an old stump, each clump of two or three stalks anchored at the base by a tangle of branched, rootlike strands of mycelium. As the spore-bearing stalks approach their predestined height, from 1/10 to 1/8 inch, the tips expand into spheres, filled with protoplasm and nuclei. Within hours a cell wall forms around each nucleus in this head, thus forming the spores. The whole mass of spores within each head or spore case — technically called a sporangium — is surrounded by a thin wall. As the spores ripen, this wall dries out, breaks, and scatters the spores to the wind.

The sporangia of Rhizopus are borne in such profusion that a culture on agar or bread or bananas or potatoes will be black with them, each head containing upwards of fifty thousand spores. Starting from a single spore, this fungus can produce hundreds of millions of spores within three or four days, each of them ready to go out and make its moldy way in the world. Rhizopus always is growing vigorously on the mantle of plant debris that covers most of the earth; it can be found in any yard or gutter as well as on plant materials on farms or in forests. It is a cosmopolitan plant, as are so many other of the common fungi that convert dead things into living mold. The speed with which these fungi grow and the number of offspring they produce help to explain why the air always carries a heavy, if invisible, freight of fungus spores, and why anything exposed to the air for a few seconds is almost certain to be peppered with the spores of different kinds of fungi.

Many fungi produce two or more kinds of spores or produce their spores in different ways from those described above. A few deserve a brief description here as examples of the variety encountered in spore production.

Endothia parasitica causes chestnut blight. It was introduced into the United States around 1900, and in the next forty years eliminated almost 100 per cent of the commercially valuable stands of chestnut in the country. Its phenomenal success is due partly to the number of spores it produces and to the variety of ways in which these are spread.

The mycelium of Endothia invades the bark of chestnut trees through small wounds, such as those made by the claws of squirrels or woodpeckers, grows through, kills, and digests the bark, and within a couple of weeks begins to produce tiny, pimple-like fruit bodies just beneath the surface of the bark. These begin as solid clumps of mycelium. Each clump or knot of mycelium grows until it ruptures the outer layers of bark and so is exposed to the air. As this clump of mycelium, destined to become a fruit body, enlarges, a cavity is formed within it by partial digestion of the cells composing it, and a narrow channel is formed to connect this chamber with the exposed surface. Stalks grow out from the walls of this cavity, a spore is formed on the tip of each stalk, released as soon as it is mature, another is formed, and so on, until the cavity is filled with millions of spores, all of them embedded in a sticky matrix. This matrix has the ability to absorb water and swell rapidly. When moistened by rain, it swells; spores and matrix are forced out of the fruit body in the form of a twisted red or yellow tendril. A single one of these small, pimple-like fruit bodies, technically known as a pycnidium, may exude from half a billion to a billion spores. It is not unusual for at least fifty of these fruit bodies to be produced per square inch of bark. On a heavily infected tree there are literally dozens of square feet of bark densely populated with such fruit bodies. The number of spores produced is incalculable.

When moist, the matrix in which the spores are embedded is sticky. Insects, birds, and squirrels that come in contact with it pick up large quantities of spores, and if they visit an uninfected tree may carry them along and inoculate the new tree. Nearly a billion of these spores have been washed from the feet of a single woodpecker. This particular one of our feathered friends undoubtedly helped spread chestnut blight up and down the Appalachians during fall and spring migrations and helped spread it locally during the summer.

The fungus does not depend on these sticky spores alone for its spread. A few weeks after pycnidia have begun to exude their spores, another type of fruit body is formed deeper in the bark. These are in the shape of hollow spheres, with necks reaching up to the surface. Several dozen of these spherical fruit bodies, known as perithecia, may be formed beneath and around a single pycnidium. From the inner wall of each perithecium, sacklike cells grow out. These are known as asci. Eight spores, called ascospores, are formed within each ascus. Some hundreds of thousands of asci are formed within each perithecium. When the eight ascospores are ripe, the ascus in which they are enclosed is released from the wall of the perithecium, travels up the neck to the opening, and there explodes, shooting the spores out into the air. The whole process is admirably adapted to getting these spores out into the free air, where they can be carried by the wind.

Thus the fungus produces two different kinds of spores, adapted to different means of transport. One is carried principally by birds, rodents, insects, and splashing rain water, the other chiefly by air currents. Both kinds are produced in astronomical numbers. Both endure drying, freezing, and other brutal influences without loss of vigor. No wonder that once the fungus became established on a few trees in this country, it was impossible to eradicate it. There are some thousands of fungi that produce spores similar to those of *Endothia parasitica* and that follow a similar pattern of life.

The mushrooms, or gilled fungi, of field and forest, and the bracket fungi common on decaying wood are also prolific producers of spores. In fact the sole function of the mushroom or of the woody bracket is to produce a crop of spores. The gilled fungi produce their spores on the sides of gills, and the pored fungi, which include most of the woody brackets, produce them on the inner walls of pores. In both, the spores are produced in groups of four on the tips of club-shaped cells known as basidia. The spores themselves are known as basidiospores. These basidiospores do not merely drop off their stalks when ripe; they are shot off, with just enough force to carry them out into the open space between the gills or into the middle of the pore; then they fall until they are free of the fruit body, and are carried away by the wind. The fruit bodies of these fungi are remarkably efficient spore-producing devices, and a whole series of complex and delicate adjustments have been evolved to insure production of the largest number of spores in the shortest possible time and with the least possible confusion. It is not a haphazard process, but rather one of meticulous order and precision, the complex product of countless ages of evolution.

While the subject cannot be gone into here, it may be mentioned that many fungi, including some of those described above, produce both asexual spores and sexual spores. Some aspects of the sex life of the fungi will be taken up later on, not only because it is an important part of the biology of molds but also because it gives us more basic information on the biology of sex than we are likely to get, say, from the study of sex in human beings. The fungi have explored many different possibilities in this field and have evolved some rather clever and effective variants of the standard boy-meets-girl theme.

A few of the fungi do not produce spores at all — or at least are not known to produce spores, which is quite a different thing. What is more of an anomaly than this is that some fungi produce spores which apparently are functionless. The fungi that combine with algae to form lichens often produce large numbers of ascospores, which are shot out and spread by the wind. So far as anyone knows, these spores are useless, because the lichen can grow only when fungus and alga, the two partners, are spread together. This spread is accomplished by the formation of small, powdery bits of mycelium in which a few algal cells are entangled, the whole easily spread by wind, insects, birds, and other means. The spores, into the production of which a considerable part of the energy of the lichen must sometimes go, appear to be a hangover from the past — morphologically prominent but functionally vestigial. Similarly, the cotton-root rot fungus, which has been a curse to agriculture in portions of the southwestern United States, regularly produces asexual spores by the billions. So far as known, these spores never germinate. In the economy of the fungus they seem to be useless, a waste of energy. There is an old theory in biology that every development, every structure, in living things has some essential function. While this theory may be useful to us biologists in lecturing to our students, it seems doubtful whether some of the fungi are aware of it. It probably is just another scientific myth.

In summary, fungi are trying different methods and means of reproduction. If we were to judge evolutionary success by ability to survive and take over the earth rather than by morphological complexity, they would as a group rate a fairly high position.

The size and shape of fungus spores. The smallest fungus spores are about a micron in diameter, no larger than the biggest bacteria. The largest are about 300 microns long, just big enough to be seen with the naked eye. The majority of fungus spores range between 3 and 30 microns in diameter or in greatest length. As with other microscopic forms of life, significance is not correlated with size.

Fungus spores vary in shape from spherical and oval through crescent and star-shaped. Some are coiled like spiral springs. Others are beautifully ornamented with spines, warts, bumps, or ridges. Some are whiskered. A few of the spores any student of fungi encounters in his everyday work are illustrated in Figure 4.

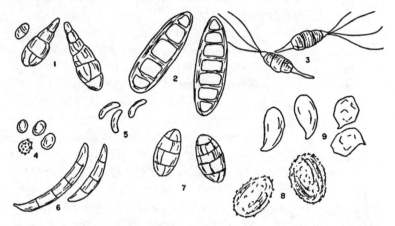

Figure 4. Spores of some common fungi. (1) Alternaria. (2) Helminthosporium. (3) Pestalozzia. (4) Aspergillus and Penicillium. (5) Cytospora. (6) Fusarium. (7) Ascospores of Pleospora. (8) Urediospores of a rust fungus. (9) Basidiospores of mushrooms.

Some of the spores and spore-bearing structures have a grace and beauty seldom encountered in the larger forms of life, and together with other microscopic plants and animals they would bear exploiting in the decorative arts.

The numbers of spores produced by fungi. Some hints of the fecundity of fungi already have been given, but additional evidence seems worth presenting because this fecundity is one of the major factors that make the world as moldy as it is.

Ustilago zeae, or *U. maydis,* causes smut of corn, and has been a minor or major pest of this plant ever since the Mayan Indians in ancient times developed corn into a cultivated crop. The fungus grows in the stems, ears, tassels, and indeed any and all above-ground parts of the corn plant, producing galls of various shapes and sizes. Each gall when mature is a solid mass of black spores of *Ustilago zeae*. Each spore in this gall is about 10 microns, 1/2,500 inch, in diameter. A single cubic inch of smut gall contains, in round numbers, 6 billion of these smut spores. A gall of average size, usually several cubic inches in volume, contains around 25 billion spores.

An acre of corn, with galls on 10 per cent of the plants, which is a very moderate degree of infection in many of the chief corn-growing regions of the world, would thus produce about 50 trillion smut spores. Some years there are at least several times that number on an average acre of field corn. In the United States we grow about 100 million acres of corn every year, not counting the innumerable small garden plots, each of which also contributes its share of spores to the total. If all the smut spores produced on just a single average acre of field corn got into the air at one time, they would furnish 10 spores per cubic foot for 34 cubic miles of air. If all the spores from 100 million acres got into the air at the same time, they would contaminate 3.4 billion cubic miles of air with 10 spores per cubic foot, or 10 spores per cubic foot in the air a mile deep over an area of 3.4 billion square miles.

Not all of these corn smut spores do get into the air at one time, of course, and many of them never get into the air at all, since they may be washed to the ground by rain, put into silos to be eaten by stock, or remain in the galls on standing plants. But enough of them do get into the air so that most of us inevitably eat, drink, and inhale a fair quantity of them winter and summer.

Tilletia tritici, which causes bunt, or stinking smut of wheat, is an ancient and shifty enemy of this important crop, and a prolific one. In a wheat field with only 1 per cent of the plants infected, this fungus will produce about 5 billion spores per acre. Infections of 10 to 30 per cent of the plants are not at all uncommon on some varieties of wheat in some regions. When wheat heavily infected with this smut is threshed or combined, the powdery spores are liberated in dark, visible, and musty clouds. In the past they sometimes were responsible for dust explosions around threshing machines, the spore clouds being ignited by sparks of static electricity from the machinery. Devices were installed on threshing machines to ground static electricity and so prevent such explosions. Showers of spores of this fungus have been caught more than a hundred miles from the nearest possible source, and it is likely that they drift much farther than that.

A very moderate degree of infection of stem rust of wheat, caused by the fungus *Puccinia graminis*, will produce a crop of up to 10 billion spores per acre. In a heavy rust year on the Great Plains there may be hundreds of square miles, with 640 acres to the square mile, of wheat with a much more than moderate infection. These spores are easily caught up by the wind and carried for hundreds of miles. Spores of this fungus picked up by the wind in rusted fields in Texas, Oklahoma, or Kansas may settle as an invisible shroud over the wheat fields of the northern states and the central Canadian provinces. Nearly a million spores of this fungus were once caught per square foot near Fargo, North Dakota, when there was no rust within hundreds of miles. Canadian workers have caught these spores in the air north of the Arctic Circle, hundreds of miles from any possible source.

A single fruit body of *Fomes applanatus*, a common wood-rotting fungus of the temperate regions of the world, was estimated to have liberated 5.4 trillion spores over a period of six months. This was approximately 30 billion a day, or 1.25 billion an hour, or 21 million a minute, or 350,000 a second. Day in and day out, for six months. In still air on a humid day, the spores can be seen falling from such a fruit body as a wraithlike cloud that drifts off on air currents too delicate for us to feel. A single log or stump may bear half a dozen or more fruit bodies of this common wood-rotting fungus. In a hardwood forest with many fallen trees and old logs about, there may be at least several hundred fruit bodies of *Fomes applanatus* per square mile. Even if each of them produced only one tenth as many spores as the one mentioned above, this would still mean a great many spores of *Fomes applanatus* in the air.

There are some hundreds of different fungi that grow on and rot wood and that produce tremendous numbers of spores. So many spores that wood exposed to air, inside or out, will inevitably become inoculated with the spores of some of the fungi that decay wood. If the wood is moist, these fungi will grow and cause decay. We have exposed sterile wood blocks to the air in late winter and early spring, then put the blocks into sterile jars and left them for a time. Blocks exposed to the air as long as twenty-four hours, even in winter in Minnesota, almost invariably became decayed — proof that spores of these fungi were present in the air. In summer the air probably is filled with them.

Mushrooms, too, are prolific spore producers, and it is not unusual for a single fruit body of a mushroom to liberate some hundreds of millions of spores in a single night. Puffballs, which are relatives of the mushrooms and wood-rotting fungi, at maturity consist of a mass of powdery spores. In a giant puffball, more than a foot across, there are billions of billions of spores. One of the smaller puffballs, *Lycoperdon pyriforme*, is common on decaying wood in the forest. At maturity it consists of a pear-shaped sack, about an inch in diameter, with a pore at the top through which the spores can escape. The fungus has no means of discharging its spores forcibly, but the wall of the puffball is so constructed that when raindrops fall upon it, it gives a bit, and the pressure of a single raindrop will blow a million spores out of the fruit body.

The foregoing has given some idea of the number of spores produced by fungi. It should also suggest that many of the fungi are rather complex, highly developed organisms, with a multitude of delicate and accurate adaptations to ensure survival. The toadstool you tread underfoot is not just a primitive excrescence defacing the lawn: it is a marvelously refined spore-producing apparatus, both mechanically and physiologically. An acquaintance with some of its subtleties helps us to realize that our modern civilization is not necessarily the final, culminating product of organic evolution.

How Fungus Spores Are Disseminated

Because fungus spores are so small and have so large a surface for their mass, they fall with surprising slowness in still air. By still air is meant the air inside a closed tube in a uniformly lighted and heated room. The spores are so buoyant that a beam of light projected through the tube in which they are falling will create air currents strong enough to send them churning up like smoke from a puffing locomotive. In nature, the spores liberated from a mushroom or a shelf fungus often will "fall" upwards and form visible deposits on top of the fruit body. In one test a mushroom was placed at one end of a horizontal closed box 10 inches high and over a yard long. Some of the spores liberated by the mushroom in the supposedly still air within the closed box floated almost to the opposite end, nearly a yard away, before they got to the bottom.

The rate of fall of a number of fungus spores has been measured in still air. The largest of these spores, nearly 80 microns long and 25 microns wide, falls at the dizzy rate of a foot in half a minute. Some of the smaller ones, less than 5 microns long, fall at the rate of a foot in from five to thirty minutes. The rate of fall in still air of spores of a few of the common fungi and the calculated distances to which they could be carried by a twenty-mile-an-hour wind, from an altitude of one mile, are given in the following table:

	Rate of Fall	Time Required to Fall 100 Feet	Miles Carried
Alternaria	3 mm/sec	2½ hours	2,900
Helminthosporium	20 mm/sec	25 minutes	440
Puccinia graminis .	12 mm/sec	42 minutes	740
Ustilago zeae	3½ mm/sec	2⅔ hours	2,500

The calculation of the distance to which such spores, once they had got a mile above the surface of the earth, might be carried by a wind of a given velocity, is, of course, merely an exercise in logic and mathematics. It has little connection with biology. We know that air currents regularly carry the spores of many kinds of fungi up to an altitude of several miles. Living spores of several different kinds of fungi have been caught more than seven miles above the surface of the earth. If air currents carry these spores up several miles, the convection currents normally moving in any air mass are likely to keep some of the spores suspended almost indefinitely. If the air mass comes down, the spores come down with it. If the air mass moves along at a more or less constant altitude, as some air masses do, the spores will be carried with it. Raindrops carry down spores from the air: according to limited tests we have made, the first raindrops of a passing shower may be heavily laden with fungus spores, but as the rain continues, fewer and fewer spores are found in the droplets.

By using a "tracer" or "marker" fungus, one whose spores are not normally present in the air, we have traced the dissemination of spores through buildings. Spores liberated in a room on the first floor of a four-story building were caught in rooms and hallways of the second, third, and fourth floors within five minutes after they had been released. In the second five-minute period after they had been released on the first floor, they were caught by the hundreds per square foot on the third and fourth floors. This probably meant that some thousands of spores per cubic foot of air were present on the upper floors.

Air-borne contamination by fungus spores is often a problem of some importance in bakeries and other food-processing plants. Sometimes attempts have been made to overcome such contamination by maintaining clean, sanitary, and relatively mold-free conditions in the rooms where bread is cooled and wrapped, where macaroni or spaghetti are conditioned, or where butter is packaged. Such sanitation is essential, but it is not the only essential, as the above tests show, since the source of contaminating mold spores may be several rooms or floors away. When such mold problems occur in industrial plants, it is necessary to track down the source of the mold and eliminate the source. A fungicidal paint on the walls of the room where the bread is cooled is of little value if most of the spores that contaminate the bread are coming from a basement room where moldy bread or other moldy materials are allowed to accumulate.

The work on the dissemination of fungus spores within buildings also suggests that someone at least four floors below us can sneeze in our face rather effectively. It may also help to explain why some diseases caused by bacteria sometimes spread rapidly through a given hospital. Some of the answers to certain private and public health problems can be found only by tracing accurately fungus spores or bacteria carried by the air. In the field of plant pathology — the study of plant diseases — the spread of air-borne inoculum has been rather intensively studied because it is of such great importance. Many of our most destructive plant diseases are spread chiefly by air-borne spores. A few cases will be described.

The tremendous number of spores produced by *Puccinia graminis tritici,* which causes stem rust of wheat, already has been mentioned. These spores, known as urediospores, are produced on wheat plants and can infect other wheat plants. These air-borne urediospores are known to be the major factor in the spread of rust northward from northern Mexico and southern Texas over the Great Plains and into Canada every year. In the fall they are often carried southward and infect the young winter wheat just coming up in the southern part of the wheat-growing region. Nearly thirty years of detailed studies of the air-borne spread of this number-one enemy of wheat have served to trace epidemic after epidemic from south to north.

The elimination of barberry, the alternate host of this parasite, was the first big step in licking the rust. The next big step was to develop rust-resistant varieties of wheat in the north, varieties that would not be infected by spores blown up from the south. The next big one, now going on, is to develop rust-resistant varieties of winter wheats in the south, so that little rust will develop on them; few spores will then be carried up to infect northern wheats, and southern wheats will not be infected by spores from the north. This step has progressed far enough so that we are well on the way to eliminating stem rust as one of the major pests of wheat, for the first time in the history of agriculture. Well on the way. We haven't yet arrived. Stem rust is another of the shifty enemies of our agriculture that will not be entirely eliminated for a long time to come, if ever. One of its most effective weapons has been the production of large numbers of urediospores that can be effectively spread by the wind, but it has other weapons too.

It is not only in the United States that the spread of urediospores of stem rust has been a major factor in agriculture. They spread from the plains to the mountains of India in the hot season and back down to the plains in the season when wheat is grown. They spread vast distances in Russia. They are supposed to have spread by air from Australia to New Zealand, although this is only conjecture. They probably spread across the Mediterranean from Africa to Italy. The lack of general north-south winds in Europe, because of the east-west chain of mountains in Switzerland and southern Germany, has kept stem rust a minor problem in France, Germany, England, and the Scandinavian countries ever since barberries were eradicated there.

Even a bare acquaintance with some of the aerobiological problems connected with stem rust is likely to give one's thought an international cast. And it is only fair to state that some of the most productive work on this problem, as well as some of the most productive thinking along international lines, has come out of the traditionally isolationist Midwest. Provincials in the world's capitals, especially those in government, should at least become familiar with some of the aspects of this problem, for it still is important to our survival and prosperity.

Stem rust is not the only fungus that spreads by air. About seven hundred species of smut fungi, most of them parasitic on members of the grass family, many of them serious pests of our cultivated cereals, depend principally upon air currents to disseminate their spores. The downy and powdery mildews cause destructive diseases of a great variety of cultivated plants throughout the world, and their spores are disseminated almost only by air. Man has carried, and continues to carry, many of them across oceans, from one country to another, as he has carried rusts, smuts, and other fungus pests; but air-borne spores quickly spread these diseases throughout large areas once they have been established on a few plants in a given continent. The fungus that causes late blight of potatoes may be carried from one country to another as dormant mycelium in potato tubers, but in a given region its annual spread is almost only by air-borne spores. Epidemics of late blight in Aroostook County, Maine, U.S.A., or in Kent, England, depend upon the spread of these spores by air during the growing season.

So it is with many other plant diseases caused by fungi. Airborne spores of fungi have always been a potent factor in the economy of man, for they often determine what crops can be grown where, and thus have been a factor in the rise and fall of various civilizations. Though faith, logic, and cloudy psychological influences may help to shape our actions, our life still depends principally, primarily, and first upon food. Even a philosopher or a historian must eat. Air-borne spores of fungi often determine not only what they will eat, but how much, and how good. They live by bread, as do the rest of us, and they do not relish moldy bread any more than a mycologist does.

Not all fungus spores are spread by the wind. Some fungi have developed spores that are carried around principally by insects. A large number of fungi are carried almost only by insects. Not by chance, but by design on the part of the fungus. The sticky conidia of the fungus that causes chestnut blight already have been described. They can be picked up and spread by insects as well as by birds and rodents, but they are not specially adapted to insect dissemination. The spores of some fungi are.

Ceratostomella is a genus of fungus that produces ascospores in fruit bodies with long beaks. The fruit bodies are called perithecia. The beaks of these perithecia may be as much as a millimeter long. The perithecia commonly are formed on the walls of beetle tunnels, the beaks protruding out into the tunnel. The ascospores are not shot out, as are the ascospores of most fungi. The ascus walls dissolve into a water-absorbent colloid within the perithecium, this swells when wet, and sticky matter and spores are exuded from the tip of the long beak in the beetle tunnel.

The exuded spore mass is in an excellent position to be picked up on the hairy outside of passing beetles, and also to be eaten by them. The fungus can be spread either by being rubbed off the outside of the beetle when it enters a new tree or by growing from the excrement of the beetle. The species of Ceratostomella that causes blue stain of wood as well as those that cause diseases of living trees, including the much publicized Dutch elm disease, are spread in this fashion. Fungus and insect are interdependent. Over the hundreds of thousands of years that they have been working together, both have evolved adaptations that make for the survival of the partnership. In this evolution they have come to depend almost absolutely upon each other: together they succeed, alone they could no longer compete. The ambrosia beetles and the leaf-cutting ants are entirely dependent on the fungi they cultivate, but this will be gone into in another chapter.

Most of the rust fungi produce, at one stage in their life cycle, pycniospores that function as gametes. These spores must be carried from one pycnium to another to ensure the fertilization of the fungus and to ensure the production of the aeciospores that follow in the cycle. Many rusts, when producing pycnia, induce a yellow color in the host tissue and exude a brightly colored, sweet fluid. The bright yellow color is one that can be seen easily and from afar by flies. Attracted by it at a distance, they are enticed by the odor when they come closer, and they wallow in the spores. These they transport from one pycnium to another, and so ensure the fertilization of the rust fungus even as other insects ensure the fertilization of flowers. Although the rust fungus may be relatively simple in structure, in the matter of insect attraction it is as complex as some of the higher flowers.

Stinkhorns are obnoxious, if harmless, fungi that grow in decaying wood in lawns. They send up thick stalks, surmounted by a sticky, ill-smelling mass of spores. The odor, so unpleasant to us, is attractive to carrion flies, who often cluster in hordes around a freshly risen stinkhorn. They mistake its odor for the odor of carrion. They mess around in the evil liquid, pick up a heavy load of spores, and presumably carry them to those places where stinkhorn spores can grow. So far as is known, the flies get no benefit from this. But it is somewhat remarkable that an organism at the bottom of the scale of evolution, as fungi supposedly are, should have found out that certain kinds of flies relish the odor of carrion, have then gone ahead and produced it, together with a sticky material in which its spores are embedded, and finally have developed a means of getting the ill-smelling mass up into the air where the flies can get at it. Our common stinkhorns appear to be rather complex organisms. Certainly they are not just the crude and random excrescences of the soil that they are likely to be thought by the ordinary citizen. They are one of God's ill-smelling wonders.

Truffles and some of their relatives, which produce their spores beneath the surface of the ground, can be spread about almost only by insects, rodents, pigs, and other animals attracted to them. To this end the truffle of commerce has produced an odor attractive not only to some beetles and to pigs but also to man. Its odor and flavor are the essence of a truffle, so far as a pig or a beetle or a man is concerned. To the truffle it is only a way of getting its spores spread far and wide. There are many kinds of truffles. Only a few of them are attractive to man. It may be significant that these have developed in those regions where the Cromagnon man, ancestor of at least some modern Europeans, lived, fought, loved, and died. Perhaps the truffle is one fungus that made at least a start at a partnership with man. Or perhaps man just happens to share with pigs a love of truffles. Truffles will be gone into more completely in the next chapter. At present the main point is that they have developed means which ensure the spread of their spores by insects, as well as by pigs and man. And they have evolved some rather subtle biochemistry to promote their ends.

Additional Reading

American Association for the Advancement of Science, *Aerobiology*. Publication No. 17. 1942. 289 pp.

J. G. Leach, *Insect Transmission of Plant Diseases*. New York: McGraw-Hill Book Company, 1940. 615 pp.

F. C. Stakman and Clyde M. Christensen, "Aerobiology in Relation to Plant Diseases." *Botanical Review*, 12: 205–253 (1946).

CHAPTER 3

Fungus Partnerships with Other Plants

FUNGI have formed all sorts of partnerships with other plants. Some of these appear to be fairly loose associations, each partner being able to survive as an individual, rugged or otherwise. Others are obligatory in the sense that each member of the combination is totally dependent upon the other, each producing certain goods or services necessary for the survival of the team. This happy union, in which each member complements and benefits the other, grades by insensibly small steps into a condition where either the fungus or its associate is little better than a slave to the other. In almost all of these obligate partnerships it would seem that one or both members have attained a degree of security and stability that they could not have attained alone. For a promise or the fact of survival, one or both members have sacrificed independence.

It is not intended to force analogies between these associations and those which may occur in other plant and animal societies, including our own, but some comparisons are almost unavoidable. Fungi have had a far longer space of time than we have had to explore the possibilities involved in partnerships. Some of their cooperative associations were going concerns at least 350 million years ago, long before man was even a gleam in the cosmic eye. They afford us an insight into biological teamwork and the subtle exploitation of one partner by another, and so it is both interesting and instructive to see what they have done, and who has done whom.

Lichens

Economic aspects of lichens. Lichens almost literally cover the earth, but so humble and inconspicuous are they that even those who pursue research solely for its or their own sake have not devoted much time to them. Explorers in the arctic and antarctic regions occasionally mention them, but only in passing, and then only because lichens are almost the only green plants to be seen in certain parts of those forbidding regions. Lichens may have had a good deal to do with making the world as it is today in a biological sense, but they do not partake of the heroic or dramatic. Yet to a botanist or mycologist they are often fascinating.

A few lichens have been used as a source of dyes of one kind or another. Harris tweeds, if they are genuine, are still made from yarn dyed with a lichen soaked in human urine. During the late Middle Ages there was a fairly extensive European traffic in certain of the lichen dyes, with some small fortunes made and lost. Litmus, which becomes red when acid, blue when alkaline, comes from a lichen, and was long used as a crude indicator of acidity.

Lichens have at times been used for human food — the Biblical manna is supposed by some to have been a lichen that was carried by the wind from its place of growth, and gathered by the hungry people where it fell. But most lichens are bitter in flavor and relatively unnourishing. If one were lost in the woods or in the far north, it might be worth while to know that certain lichens are edible, after a fashion; but it would be much better to have some pemmican or other concentrated food along. Reindeer and some other wild animals subsist mainly on reindeer moss and other lichens during the winter. In some northern regions there are many square miles of lowland covered almost only with reindeer moss, *Cladonia rangiferina,* one of the larger lichens. Stretches of it acres in extent can be seen in the northern United States and southern Canada.

Over a long period of time lichens have been of considerable biological importance. They grow chiefly upon rocks, soil, tree bark, and wood. They are almost the only plants that grow upon bare and barren rock, certainly the only visible ones, and they slowly convert the rock into soil. They literally decay it. Rocks may be broken down into soil by many different means. Volcanic action, abrasion, chemical deterioration, and the action of fungi, bacteria, and other microorganisms may be involved in the conversion of stone into rich soil. Originally the earth supposedly consisted mainly of rocks and water. The lichens have had much to do with converting the original rock into soil, and presumably they will continue this work in the future. It is not exactly dramatic work, and it is infinitely slow, but merely because it is not suited to presentation by radio or television does not mean that it is of no significance.

A few lichens cause plant diseases in the tropics, and as agriculture becomes more intensive there, lichens may become a greater factor in the economy than they now are. From this purely practical standpoint they would bear very thorough investigations in certain regions.

What lichens are. Each lichen is a combination of a fungus and an alga. Taxonomists have had some bitter arguments whether lichens should be classified with the algae, with the fungi, or put in a special group by themselves. The last possibility was especially distasteful to those who held with almost fanatic tenacity to the view that there is a special niche for every living thing in the general scheme of evolution, and that it constituted rank apostasy to classify *combinations* of two quite different kinds of organisms in a separate group of their own. However, if lichens are combinations, as they are, it seems better to recognize it, and to classify them as combinations. This is what has been done by all except a few diehards who insist upon fitting nature to logic, and now lichens are mostly classified as lichens. Even in taxonomy, common sense sometimes eventually prevails.

The fungus partner makes up the greater part of most (but not all) lichens. The fungus not only makes up the bulk of the lichen, but determines the form of the plant, its shape, size, and manner and place of growth. The fungus mycelium penetrates the substrate on which the lichen grows—rock, tree bark, wood, or soil—and secretes acids and enzymes that break down the substrate for food. The surface layers of the lichen are commonly tightly woven and partly gelatinized fungus mycelium, designed to take up water rapidly and lose it slowly. The algal cells usually are entrapped among a loose tangle of fungus mycelium just beneath the lichen's surface, almost as if held in bondage by the fungus. In many cases they doubtless are. (See Plate 3.)

In some lichens the fungus mycelium is merely in close contact with the imprisoned algal cells. In others the fungus mycelium penetrates into the interior of the algal cells. The algal cells apparently get raw materials from the fungus partner, and with the aid of such sunlight as penetrates the wall of their fungus prison, transform these materials into a great variety of more complex foods. Some of these are essential to the growth of the fungus partner, or fungus master.

The algal partner in many of the common lichens is identical with, or greatly similar to, certain independently living algae in fresh water ponds and lakes. The fact that they are structurally similar to free living algae does not mean that they are physiologically identical with the free living forms, or that they could live an independent life. The algal partner of a few lichens has supposedly been separated from the fungus and grown as an independent plant in the laboratory. Overlooking for the moment some of the technical qualifications that may be attached to this work, it looks as if the algal partner, where the alga can be grown independently, is less of a partner than a wild plant that has been more or less domesticated by the fungus.

In other lichens the alga is the dominant partner, making up most of the plant and determining how and where it will grow. In these cases it looks as if the fungus were being exploited. In many, perhaps most lichens, neither the fungus nor the alga can be grown alone; each is essential to the other. Each has won security and survival, but lost freedom of choice.

Lichens are apt to be rather leisurely in their growth and expansion. An ankle-high growth of reindeer moss may be centuries old. A patch of green or yellow lichen on a rock may be a centenarian, and many of the lichens on the bark of trees grow for at least decades if not centuries. On the other hand, I once encountered a lichen which grew with such vigor on a flat asphalt roof, the roof being constantly sprinkled with water to keep the rooms below cool, that it covered the roof with a thick mat in two short summers. When it dried up during the second autumn, after the roof no longer was watered, it curled up and pulled large patches of asphalt and roofing paper with it. It ruined the roof, and incidentally caused some interesting (to an outsider) legal contention between those who made the roof and those who had paid for it. When grown in the greenhouse, this lichen developed from a number of small fragments to a thick mantle which covered an asphalt and gravel surface ten inches in diameter in a few weeks. All on such nourishment as it could get from distilled water, asphalt, gravel, and dust from the air.

Few of the lichens are dominant plants except in very specific and limited locations. They grow almost only where nothing else can or will grow. They digest stones, tree bark, or pieces of wood too dry for wood-rotting fungi to occupy, or thrive in swamps in certain periods of the evolution of swamps to dry land. By means of the partnership between fungus and alga they can exploit these unpromising environments, in which competition is nearly nonexistent. By virtue of it they can grow where neither of the individuals in the association could grow alone. As a team they seem to have assured themselves of a variety of places in the sun as well as in the shade. And also to have assured themselves of long-time survival. If in some cases one or the other or both have given up independence for the sake of permanence, this is just one of the risks of partnership. The statement "He who has partners, has masters" is well exemplified by the lichens. Fungi and algae had explored this truth long before humankind.

Even in reproduction of the dual plant, the two partners work together. In many lichens the fungus partner, especially where it is the dominant one, will produce large and conspicuous fruit bodies. Since most of the fungi involved in lichens are ascomycetes, the spores formed in these fruit bodies are ascospores. In some cases a considerable portion of the .itality of the lichen must go into their production. Yet, as has already been said in the discussion of functionless spores in the preceding chapter, these ascospores seem to be mostly worthless in reproducing the lichen and are simply wasted. The lichen as such, the combination of fungus and alga, reproduces itself by forming tiny clumps in which a few algal cells are surrounded by a tangle of fungus mycelium. These minute reproductive units, formed in large numbers on the surface of the lichen, are blown or carried about to reproduce the plant. Typical asexual reproduction, with the added twist that here both of the partners have to be intermingled in a single reproductive unit. Judging by the prevalence and distribution of lichens, the method is a very successful one.

No one knows when or how the fungi and algae first struck up an acquaintance that finally ripened into this closely knit communal life. Perhaps they were associated casually in the sea when the world was young and before land plants had evolved. As the mountains heaved up and life on land became possible, the fungus and alga may have combined by chance into a partnership that enabled them to crawl up out of the sea, invade the barren rock, and spread over the uninhabited world, slowly converting it into soil for the use of the multitude of living things that followed them. For this there is no evidence, except that we know that both fungi and algae are of ancient origin. Some lichenologists claim that lichens have evolved relatively recently, but there is no evidence for that either. We can find at present what appear to be just casual associations between certain fungi and certain algae. For example, Clavaria grows only on rotten logs covered with a layer of a certain kind of alga; certain kinds of algae grow abundantly on the surface of the fruit bodies of some wood-rotting fungi. They may be experimenting with a partnership. If it takes a hundred million years or so for them to determine if the partnership is going to work out, it does not matter. They have time.

Yet, from another standpoint, fungi and other simple organisms may seem to be rushing nowhere just as fast, frantically, and aimlessly as we do. In this running hither and yon, which an ecologist might describe as an exploration of the possibilities in their environment, they encounter a good deal of competition. When they can't eliminate a competitor, they may make what seems to be a casual working agreement with him, then form a close and tight partnership, and then exploit him for their own good. Biology is filled with examples of such cooperation and exploitation. The more one studies biology, in fact, the more one realizes that, in the plant and animal world, at least, a rugged individualist is just a theoretical abstraction.

Mycorrhizas

What mycorrhizas are. Mycorrhiza means *fungus root*, and it is a good name, because that is exactly what it is — a structure formed by the combination of fungus mycelium with the small but all important food-getting roots of higher plants. The mycorrhiza is not the fungus alone, nor the root alone, but is the combination of fungus and root. The two make a working unit, even as the lichen about which you just read is a combination of fungus and alga.

Few people ever see the absorbing roots of plants, but these roots are just as important as the leaves. They absorb water and minerals from the soil, materials essential to the growth of the plant. In most of our common annual plants and many of our perennials the absorbing roots are of the kind commonly shown in most botany texts. A germinating seed of wheat or corn forms roots that are equipped with a solid, pointed tip that enables it to push between particles of soil. Back of this armored spearhead is a region of rapidly growing cells, the region of elongation. This furnishes the necessary push to force the root cap through the soil. Back of this is the food-getting portion, where the surface cells of the root sprout out into delicate, one-celled hairs that fasten themselves onto soil particles and absorb water and dissolved minerals and a few other vital things. A young radish plant will have hundreds of thousands or millions of such tiny absorbing roots, all covered with root hairs. You do not see them because when you pull up the plant you break them off. A vigorous mustard plant a few weeks old will have a branched root system a couple of miles long, and all along this extensive system there will be uncounted multitudes of these tiny absorbing roots that gather food and water. Remove all these microscopically small, but vital, absorbing roots from any plant, and unless it can form more in a hurry it dies. That's why some kinds of plants are difficult to transplant.

Many varieties of plants, however, including a goodly number of our trees and shrubs, most kinds of orchids and their relatives, and some totally unrelated species, have few or no root hairs. Instead, the necessary drudgery of absorbing food and water from the soil is done by a fungus partner, which has sometimes become a fungus master. This partnership has had a profound effect not only on the physiology of the higher plant, but often upon its structure too. Some of the higher plants involved in this kind of partnership have become so dependent upon the fungus on their roots that they have lost their chlorophyll. Even their organic carbons must be furnished by the fungus. What started out as an adventure in cooperation has, in some cases, evolved into a dependence so complete and inflexible, and sometimes of so low a survival value, that both partners seem doomed to extinction. Some, on the other hand, are very successful indeed. A few case histories will make us more familiar with what mycorrhizas are, what they do, and how they have influenced the structure and economy of some of the seed plants.

The Indian pipe. The technical name of this plant is *Monotropa uniflora*, which means *one-turned, single flower*, and — since the blooms customarily tilt downward — the name is not a bad one. The plant is pure white or pale pink and comes up in clumps of from a few to more than fifty stalks. The stalks bear only vestigial, scalelike leaves, and a single, terminal flower. The plant is totally lacking in chlorophyll, the stuff by which its green relatives live. It is a member of the large heath family, which includes such things as heather, blueberries, cranberries, Labrador tea, laurel, and rhododendron, to name a few.

Chlorophylless seedlings, commonly known as albinos, are not very unusual in corn, wheat, and barley, and are met with occasionally in most kinds of plants. Such albinos are weaklings and can survive only for a short while. When they have used up the food stored in the seed from which they arose, they are done for. They can be kept alive for a time, sometimes even to flowering, by being fed with sugar, but in nature and on their own they quietly and quickly starve.

Monotropa, however, grows, reproduces, and lives a successful life over a major part of the world. The secret of its success is in its roots. These form a dense, hard mass a couple of inches beneath the surface of the soil. This dense mass is composed of a multitude of small, branched nubbins, each scarcely an eighth of an inch long and about a millimeter in diameter. Surrounding each of these rootlets is a thick, tightly woven layer of fungus mycelium. Anyone can see this for himself by merely digging up a portion of the root mass, washing it free of adhering dirt, sectioning a rootlet with a razor blade, and looking at the section with a microscope. This fungus mycelium covers and encloses the root nubbins like a strait jacket. Branches of the mycelium penetrate into the root, mostly between the cells, and form a network of hyphae there. Some branches enter into certain root cells and furnish them food. From the outer part of the fungus mantle, mycelium extends out into the soil. Apparently the fungus gathers food in the soil and transports it to the interior of the root, where it is utilized by Monotropa. It is of interest that rhizomes of some of the pre-conifers of the Carboniferous age, about 350 million years ago, had what appears to be an almost identical association with a similar fungus, which would indicate that such partnerships are of some antiquity.

This association has its advantages and its drawbacks, for both members. By virtue of the fungus on its roots, Monotropa is able to thrive on the forest floor beneath dense stands of trees where little sunlight penetrates. We find clumps of vigorous Indian pipes in pine woods where there is not enough light coming through the canopy to support the growth of more than an occasional spindly green shrub. In such an environment the Indian pipe has almost no competition. The plant also is able to grow in many different soil and forest types, and it has a much wider latitude in this respect, and also a much wider geographical range, than most of the larger plants beneath whose shade it usually is found.

The fungus also presumably derives some benefit from the association. In most soils the competition among the many kinds of fungi, bacteria, protozoa, nematodes, insects, and other forms of life is constant and fierce. Once established on the roots of Monotropa, the fungus dominates a small sphere of influence, and holds its own in a larger sphere. In isolating fungi from the roots of Monotropa, we have encountered a much smaller number of competing fungi than we usually find on the roots of most of our common plants, indicating that the fungus partner of Monotropa is able to more than hold its own in the small region vital to it.

The fungus and seed plant function together as a unit. If one judges success on the basis of distribution, Monotropa is fairly successful. If one judges it on the basis of the size and number of plants, Monotropa is not particularly so. In survival each partner is restricted not only by its own limitations, but by those of the other as well. It is not enough that the fungus happens to become established in a patch of soil favorable to it, or that the seeds of Monotropa land in a place favorable for their germination. Presumably, seed and fungus must arrive together in a place favorable for both. Even with our most successful plants, such as some of the forest trees or some of the common weeds, the mortality among the seeds and seedlings is extremely heavy. Only a very minute proportion of the seeds ever result in mature plants. With Monotropa, the chances of successful establishment seem reduced almost to the vanishing point, and this may be one obvious reason why Monotropa occurs only in scattered clumps here and there over most of its range.

Monotropa also is restricted in its possibilities for producing new and more successful variants. Unquestionably both Monotropa and its fungus partner are producing variant types all the time, or at least producing seeds and spores, respectively, that are genetically different from their parents. But the new biotypes, or possible new biotypes, among the Monotropa seeds, have to be adapted to survival with the average type of their partner, and vice versa. This necessity must retard the evolution of the plant and keep it more or less rigidly fixed. Or at least, once the partnership was established, about the only possibility of variation was for the seed plant to become progressively more dependent upon the fungus. In this it has gone to the ultimate — it is at the end of the line.

This sort of partnership offers some interesting opportunities for pure research in plant nutrition and in evolution. Our own efforts at research have been frustrated so far by our inability to grow Monotropa. The seeds resemble those of many orchids: they are almost microscopic in size, and the living portion consists of only ten or twelve large cells. There is no organized embryo, no endosperm of reserve food. Presumably the seed soon after germination must be spoon-fed by the fungus from the roots, but to date no one has been able to grow this combination.

Fungi and orchids. Corallorhiza means *coral root,* and is the name of a genus of orchid, with about a dozen species, that occurs in north temperate climes. These orchids are not abundant: in a day's ramble through the brush and bramble where they grow, a person may find one or a few clumps. They look a good deal like a larger, more deluxe edition of Monotropa, but they suggest decadence. They look as if they were ashamed of their shiftless ways, and wanted to flower quickly and be gone. Or as if they had found the struggle tiring and futile. For them, it may be. Like Monotropa, they have no chlorophyll, and so are white or have a pale pink tinge. Their roots are a fist-sized clump of brittle, coralloid branches. Like the roots of Monotropa, these do not have a single root hair, and depend for their food upon a fungus. Unlike Monotropa, the fungus here is mainly within the root; there is no mantle of mycelium around the outside. The cells near the periphery of the roots are packed full of fungus mycelium; toward the root center the mycelium becomes less abundant. The mycelium apparently gathers food in the soil and transports this to the orchid, and within the root some of the mycelium is digested and its food utilized. In turn, the orchid apparently contributes certain substances to the fungus that it could not get otherwise. The fungus associated with Corallorhiza has not been identified and named, but it must be even less aggressive than the one working for Monotropa, since Corallorhizas are much less abundant than Indian pipes.

Corallorhizas are not the only orchids that depend for life and food upon a fungus at their roots. Most orchids do, including the elegant lady-slipper or moccasin flower of the northern regions, the showy Catteleya, and many tropical kinds that have long been cultivated in greenhouses. When tropical orchids first were cultivated, it was found necessary to plant the seed in soil taken from around established orchids of the same kinds, or, better yet, to place pieces of roots from established orchids in soil where seeds of that same kind of orchid were planted.

The reason for this practice was not known. It was followed because it worked. Some orchid growers still follow it because it works. Later it was found that the roots of most of the orchids investigated were mycorrhizal, similar in structure and economy to those of Corallorhiza described above. Once this was known, it was possible to isolate the fungus from the roots, inoculate it into sterile soil so that it was the only organism present, then plant the seeds there. This assured the germination of the seeds and the establishment of seedlings. Later investigations proved that the seeds of certain orchids could be germinated without the root fungus if they were furnished with certain sugars and growth-promoting substances. And so now some orchid seeds are germinated and the plants grown for a time, or even to maturity, on special agar preparations in flasks and dishes, much as fungi are grown in the laboratory. Even under the best of conditions they are likely to grow very slowly: it is not unusual for a seed to require from six to nine months to germinate and produce a single small leaf, and to develop for some years before producing flowers.

The fact that orchids can be raised on nutrient agar without the presence of a fungus does not negate the fact that, in nature, most of them are dependent upon their fungus partner. Their seeds will not germinate unless the fungus partner is there to furnish them food, and the plants will not become established unless the fungus is present on the roots. The orchids with green leaves presumably manufacture a good deal of their carbohydrate food, as normal green plants do, but the water and minerals from the soil, and presumably more complex but equally necessary products of other kinds, are furnished by the mycorrhizal fungus on their roots.

This partnership probably accounts in part for the peculiar distribution of many orchids – a patch of them here, a clump a few miles away, an occasional single plant miles from its fellows. Many orchids produce seeds by the millions. These seeds, like those of Monotropa, are minute in size, simple in structure, and lacking in endosperm and an organized embryo; the living portion consists of only a few cells. They can germinate only if they happen to land on soil where their associated fungus is established and ready to feed them a well-rounded diet, or at least where the fungus is present and ready to become established. It appears also that the fungus partners of most orchids are not vigorous foragers on their own, and become really well established only when the orchid is present. The chance that both orchid seed and fungus will land together more or less simultaneously in a place where both can grow must be rather rare.

Some orchids are known in North America of which only a few individuals have ever been found, and these were hundreds or thousands of miles apart. It is not climbing too far out on the limb of speculation to hazard that in such cases the partnerships of orchid and fungus have not been strikingly successful. Even in the tropics, where some orchids flourish, not all of them are by any means widely distributed. Some are scattered here and there; some occur only in a single valley and nowhere else in the world. Many of those that are so common and ever-present in the moist regions of the tropics depend only partially or not at all upon mycorrhizal fungi.

The orchids are rather highly specialized plants. The roots of many of them have fungus partners that get their food, and without these partners the seed will not germinate, nor the seedling continue to grow. The flowers of many of them are so constructed that they can be pollinated by only certain species of insects. If plants can be said to have a social organization, the orchids have it. Whither is this leading them, in an evolutionary sense? Will they, through dependence on their partners, evolve themselves out of business? Only time will tell. The ancientness of some of these partnerships speaks well for its survival value. The football coach wants size, speed, and savvy, usually in that order, but in biological evolution these count for little. Certainly some of the ties the higher plants have established with fungi have increased the chances of survival of the higher plant. In others the higher plant that once took on a silent fungus partner to help it compete in the struggle for existence seems headed slowly but surely toward total dependence, and then extinction, the victim of its partner.

Fungi and forest trees. The study of fungi on the roots of forest trees and of the relation of mycorrhizas to the trees' growth and well-being all began with truffles. Truffles are underground fruit bodies, or spore-producing bodies, of certain Ascomycetes mainly in the order Tuberales. The truffle of commerce is only a mass of ascospores, mycelium, and delectable odor, surrounded and protected by a thick and knobby rind of mycelium. There are numerous kinds or species of truffles, but only a few of them are good to eat. More than forty different species have been described in the United States, but none of them are of the choice edible kinds. These good ones grow in various places in

Europe, but the biggest and most luscious kinds are found principally in southern France and northern Italy. They were known and highly prized by upper-crust Romans two thousand years ago, and pigs both wild and tame have been relishing them ever since there were both pigs and truffles.

Since the puffball-like fruit bodies of the truffles grow anywhere from an inch to six inches below the soil, they cannot be seen. But they can be scented, and so they are hunted by scent. Truffles have been hunted with dogs and even, so it is claimed, with goats, but the French professionals prefer pigs. In the first place, any pig has a strong inherent craving for the flavor of a good truffle and hankers after them as much as we do. Second, a pig has a keen sense of smell and can readily locate any truffle within range, needing no training to distinguish a truffle from a rabbit. In addition, a pig does not merely point at the truffle, but sets to and roots it up. The last but not least reason why the frugal Frenchman favors a pig for this delicate work is that when the animal's hunting days are over she can be served up on the table garnished with the very truffles she helped find.

The procedure of hunting truffles is about as follows: On a quiet evening of the truffle season, when the scent is likely to be most heavy and not blown about, the pig is taken under the arm or carried in a wheelbarrow to the wood where experience has shown the hunting to be good. The pig is carried or wheeled so that she may not get too tired — a good truffle-hunting pig is a wage earner and an artist, and is treated as such. Once in the woods, a rope is put around her neck, and she is given her head. When she smells a truffle she starts to root it out. She is pulled up, tied, and the hunter digs out the truffle, rewarding the pig with an acorn or some special titbit. This continues until both partners are tired or the evening's quota has been gathered.

The whole business sounds a bit fantastic, but almost the only fantastic thing connected with it is the price of the truffles. They are one of the most highly priced vegetables on the market, and have been for centuries. Many a landowner in southern France has used truffles as a good cash crop year in and year out. They have been cultivated in southern France for decades, but it is not an ordinary sort of cultivation. Oak or beech trees first are planted and, after they are established, pieces of truffle are scattered on the soil and covered. The first harvest comes after from six to ten years, and an annual crop can be gathered thereafter for two decades or more.

The odor and flavor of truffles are something out of this world, and they are every bit as attractive to most people as they are to rodents, insects, and pigs. You do not have to acquire a taste for truffles, as you do for some cheeses that are ripened — or rotted — by fungi, or as you do for "high" meat, or the fish ripened by the Eskimos. The flavor of truffles is not "high," it is altogether different: all-permeating, mouth-watering, tantalizing, with the promise of Elysian enjoyment, a condensation of all that is noble and good. One need be neither a sophisticate nor a pig to enjoy truffles.

This being the case, it is only fitting that from truffles we get much flavor and little food. Biologically, the odor and flavor of truffles are just "come-ons." Certain beetles will fly unerringly to a truffle from more than a half-mile away, by actual test. Rodents and pigs can scent them from some yards away at least. All of these animals, from insects to pigs, once they locate a truffle, burrow into it or root it out, as their nature dictates, and then gulp it down with brutish gusto. The spores, which make up most of the truffle, pass undigested and unharmed through the intestines, and are voided with the excrement. The eater gets the flavor, and some food. The truffle gets disseminated, and the spores get planted in new places, where the truffle can go about making more truffles. Fungi may be simple, but the truffle is one that uses some rather subtle and complex means to ensure that its spores, hidden in closed fruit bodies beneath the surface of the ground, will be found and, once found, effectively distributed. This is not the only remarkable aspect of the truffle.

Truffles grow only in association with certain species of trees, because the mycelium of the fungus forms mycorrhizas on the roots of these trees. The fungus mycelium apparently does a good deal of the food-getting for the trees with whose roots it is associated, and also contributes vital growth-promoting substances to the tree. In return, it gets certain substances manufactured by the tree that are essential to its growth, or at least to its reproduction. Thus we have here a sort of triple partnership, between tree, fungus, and those animals that eat the truffle and spread its spores.

Most forest trees seem to depend for a major part of their nourishment upon fungi on their absorbing roots. Thus our mighty monarchs of the woods, like other monarchs, rule with unseen help. They are far from being the sturdy individualists they seem. This partnership between tree and fungus apparently is an ancient one, and the ancient progenitors of our present conifers, such as pines, spruces, and firs, also had mycorrhizal roots and depended for their nourishment upon a fungus on their absorbing roots.

The significance of mycorrhizal fungi in the growth of forest trees has been studied for only about fifty years, and it is not remarkable that we do not yet know all the answers. It is not an easy field in which to work. The fungi involved in these mycorrhizal relationships with our forest trees are mainly what the layman calls mushrooms. Such genera of gilled fungi as Amanita, Russula, Lactarius, and among the pored fungi, Boletus, commonly form mycorrhizas. These fungi are difficult to grow on agar culture, even though they grow and fruit so prolifically in nature. Only recently has work in Sweden uncovered some of the basic relationships involved. From this research it appears that several different fungi may form mycorrhizas with the absorbing roots of a given species of tree, and that a given species of fungus may form mycorrhizas with several species of trees. Wherever the tree seedlings arise in soil that does not contain a well-rounded diet for them, mycorrhizal fungi may be essential to their growth and survival. Even when the tree is growing in rich soil, as few forest trees do, the mycorrhizal fungi may aid them. In general, about 90 per cent of the absorbing roots of a conifer tree are likely to be mycorrhizal, without root hairs, and surrounded by a mantle of fungus mycelium. It has been possible in practice to boost the growth of forest tree seedlings by inoculating the seed beds with mycorrhizal fungi. This has been especially true when the seedlings were raised in areas where no such trees occurred naturally, as was the case when nurseries were established to furnish seedlings for windbreak plantings in the Prairie Regions of the Midwest in the late thirties. That mycorrhizal fungi may be essential, or at least beneficial, to the growth of many forest trees is now becoming recognized by those who know their way around in that field, but it is by no means general knowledge. Nurserymen who grow and sell ornamental evergreens are mostly unaware of the advances in this field during the last two decades, although such knowledge might profit them greatly.

The association of fungus mycelium with the absorbing roots of forest trees is of some general biological interest. Will this Anschluss between the monarchs of the woods and the molds lead gradually to the trees' becoming more and more dependent on the fungus? The association apparently has been obligatory ever since Pre-Carboniferous times, a few hundred million years ago. It has doubtless influenced the evolution, survival, and distribution of many kinds of trees up to now, and it may be a deciding factor in their future. Botanically considered, the conifers as a group are conservative. This conservatism may be partly a result of their partnership with fungi — partnerships that for a long time have been beneficial, are now essential to survival, and later may determine even more strictly when and where the trees are to live.

Many of our conifers grow in almost pure sand, and do well in it by virtue of their fungus partner, whose wide-spreading mycelium can collect the necessary nourishment and serve it in concentrated form to the tree. From a competitive standpoint such a partnership would seem to be a good thing now. Both the trees and the fungi associated with them have sacrificed some independence and gained some security. But will the trees eventually end up as chlorophylless dwarfs like Monotropa or Corallorhiza, completely at the mercy of the humble partner they once welcomed so gladly? And, if so, does it mean that without their fungus partner they would have succumbed long ago? It looks as if these partnerships have endurance, but that once they are taken up, the seed plants have a tiger by the tail. Once committed to partnership, they cannot turn back, they cannot cancel it. Whether, among the plants, the partnerships described above or the rugged individualists, if there really are any such, will make out the best in the long run, only time will tell.

Had H. G. Wells known about mycorrhizas when he wrote *The Time Machine*, it is possible that he would have had the intrepid rider stop in a future age when the world would be inhabited only by forests of pale, succulent, watery growths that would come up, flower, produce their crop of tiny seeds, and expire, before the red and dying sun had crawled to the zenith — days or weeks by present clock time. Before the dim afternoon had well advanced, giant mushrooms would burgeon forth, overtop their once dominant partners in the expiring evening, and shower the earth with their spores through the long and silent night. Thus would biological truth be given to the Biblical contention that the meek shall inherit the earth. From a biologist's standpoint it would be fascinating to be there and see it. There is a good possibility that in the twilight of this world the forms of life predominating will be the fungi and those plants and animals smart enough to have gone into partnership with them.

Additional Reading

Bruce Fink, *The Lichens of Minnesota*. Contributions from the United States National Herbarium, Vol. 14, Pt. 1. Washington, D.C.: Government Printing Office, 1910. 269 pp.

Oskar Modess, *Zur Kenntnis der Mykorrizabildner von Kiefer und Fichte*. Symbolae Botanicae Upsaliensis, 5:1 (1941). 146 pp.

M. C. Rayner, *Trees and Toadstools*. London: Faber & Faber, 1945. 71 pp.

Edwin L. Schmidt, "Mycorrhizae and Their Relation to Forest Soils." *Soil Science*, 64:459–468 (1947).

CHAPTER 4

Fungus Partnerships with Animals

THE nature of animals is such that one would hardly expect many of them to form close partnerships with fungi. The fungi, when growing and producing food, shelter, or other forms of aid and comfort for their partners, are relatively stationary, while most animals are constantly on the go. The mobile animals, from snails to deer, will nibble at mushrooms now and then, and most of the larger animals except man regularly carry around a variety of fungi in their intestines, one of which will be described later, since it is one of the marvels of the plant world. But animals have never tried to exploit fungi in a big way. Certain insects, however, have developed a close partnership with them.

Many of the insects lead a life to which fungi are eminently adapted. Some settle down less than an inch away from their ancestral manor, and spend all of their life in that one spot, just shifting slightly this way or that, now and then, to ease their small bodies or still smaller minds. Others have family or communal dwellings, social establishments that are occupied for weeks, years, or decades by the same family or tribe, and it is not strange that some of these have found that fungi could be used for protection or cultivated for food. A few of the more well known and interesting of these partnerships will be described.

Scale Insects and Their Fungus Houses

Scale insects are small, sedentary creatures that spend most of their life in one spot on the tender bark, leaf, or fruit of a plant. They stick a long, slender sucking tube down into the plant vessels or conducting pipes, and pump out sap for food and drink. Once a young scale insect has squatted and established union with his host plant, he stays put for the rest of his life. Not being able to dodge or fight back, scale insects would not survive for long if not adequately protected. They achieve security in various ways. Most of them secrete a hard, chitinous shell or scale that covers them and the edges of which become cemented to the surface of the bark or leaf, thus sheltering them from bad weather, marauders, and most other risks. The common oyster shell scale that often literally covers the bark of ornamental shrubs and fruit trees and sucks the life out of them is a good example of this type.

Other scale insects depend upon fungi for prefabricated individual or communal dwellings that shelter them. These scale insects are abundant in the southeastern and southern parts of the United States and in most of the tropical and subtropical regions of the world. They live beneath patches of fungus on a great variety of trees and shrubs, and while what follows may be as implausible as a fairy story, it has the most excellent evidence behind it.

We'll start with a young female scale insect that has just wandered out of her maternal home. She is small, barely big enough to be seen with the naked eye, but still big enough to be considered fair game by various birds and other enemies. As she comes out into the world, she staggers by chance over the layer of fungus that protected her mother and other members of the colony, and in so doing picks up spores of the fungus. She waddles across the branch for a short distance, finds a place underneath a tiny bit of scaling bark where the future seems moderately secure, pushes her long, slender snout through the thin bark into the vital fluids of the tree, and begins to pump sap into her body. As she does this, the fungus spores she has picked up on her body germinate, send thin threads into her insides, and there grow and multiply mightily. This discommodes her only slightly, if at all. With the food so obtained from within the insect, the fungus soon produces a covering of mycelium on the outside of her body, and within a few weeks the young scale insect is hidden by a thin mantle of mold mycelium. The fungus (whose scientific name is Septobasidium) derives all of its food from the scale insect, and this particular gluttonous young female serves as little more than a pumping station to transfer sap from the tree into the fungus.

As the fungus spreads out over the surface of the bark in a patch so small it still hardly is visible, it forms not a flat growth, but on its advancing edge makes vaulted chambers, with openings toward the outer side of the patch. Other young scale insects of the same brood wander into these chambers. Some of these new members of the growing community will have been infected with the fungus before they entered the chambers, and so will contribute more food to the communal shelter, helping it grow farther and faster and form more chambers for more scale insects to inhabit. Some of the chambers will be inhabited by young females that have not been infected by the fungus before they arrived, and these remain free of infection, even though surrounded and enclosed by the fungus mat.

Those infected by the fungus do not reproduce, and the only obvious effect on them of the fungus which fills their insides is to render them sterile. They live as long, and presumably eat as much, as their noninfected sisters, but their contribution to the general good is in furnishing food for the fungus, so that the fungus can protect their fertile sisters. Those not infected eventually reproduce. Though covered and surrounded by a thick mat of fungus mycelium, there is a small trapdoor in this covering over each chamber. The trapdoor is closed by a flap of mycelium attached only at one side, and through this, at mating season, the fertile females project their posteriors to the outside, where passing males fertilize them. Thereafter these fertilized females spend their short lives, sealed in a fungus vault, guzzling sap and producing young.

The young are born practically fully formed, and after molting are ready to go out into the world. The fungus has provided for this also. There is a narrow tunnel through the mat from each maternal chamber to the outside, a tunnel formed by the fungus alone, with no help from the sedentary insect. Through this tunnel the young totter out into the world. If, after they get out, they crawl over the surface of the fungus mat that shelters the community, they may become infected with spores of the fungus. The fungus produces its spores most abundantly during moist periods in the spring, just when the first brood of young are emerging. A large percentage of this brood of insects will become infected with the fungus, while the one or two later broods will be largely free of infection. In general, around half of the young in any thriving colony will be infected, and so furnish food that enables the fungus to protect the other half of the population. Doubtless neither half knows how the other half lives.

Those individual scales not protected and housed by the fungus theoretically should be able to survive on their own, but actually they never do. They are picked off by birds and other predators, dried up by the sun, washed off by rain, or come to other untimely ends. So what it amounts to is that if these scale insects as a group are to survive, a fairly high proportion of the population must become infected by the fungus. These are sacrificed for the general good. So far as we know, it is an involuntary and quite unconscious sacrifice, and so one can hardly moralize on it very much, but it is an interesting social development.

The fungus gives the scale insects protection from more than just winter and rough weather. There is a tiny wasp that lays its eggs within the bodies of these scale insects. The wasp egg hatches into a larva, or maggot, and this larva lives on the innards of the scale, digesting and consuming the insect completely. By some radar system known only to itself, this wasp can locate the scales hidden under the fungus mat, and stick its ovipositor, or spearlike egg-laying gadget, directly into the body of the hidden scale. The ovipositor is less than a hundredth of an inch long, and thus can reach only those scales covered by a thin layer of fungus. Those covered by heavier layers are perfectly safe from this fairy-winged fiend. From the standpoint of this wasp, the chief function of the universe probably is to furnish enough scales of this particular breed, thinly enough covered by the fungus, so that it may jab its egg-laying bayonet into them. A narrow outlook, to be sure. Sometimes ours seems equally so.

There are many different kinds of scale insects that go in for this sort of partnership with fungi. All of the fungi involved are kinds related to certain common wood-rotting fungi that occur as crustlike growths on stumps, fallen trees, wood piled in our back yards, and so on. In some of these partnerships, the fungus forms ingenious traps into which the young scales blunder, and in some cases the rooms prepared for the young scales are so close fitting that once the insect has wandered in, it cannot get out, not being equipped with a reverse gear. The scale insects are not so widespread as some of their more independent relatives, and the fungus partners also are not so common as some of their independent and uninteresting relatives that live on and decay wood wherever wood is found. But the general temporary success of this sort of partnership is proved by the great abundance, in most of the warmer parts of the world, of many different kinds of these scale insects and their fungus partners on trees and woody shrubs. It is more than a dual partnership, of course, since the trees or shrubs on which both the insect and fungus subsist also enter into it, and the trees in turn may depend for food upon the molds on their roots.

Many biologists are rank individualists, and view with alarm any sort of dependence but their own. In their opinion, once an organism becomes dependent on another, either as a parasite or obligate partner, it is likely to have a rather dim future so far as evolution is concerned. They maintain that these partners and parasites tend to become more and more specialized, and eventually specialize themselves out of existence. From this they reason that, biologically, the good life is the life of complete independence. Some very convincing examples can be adduced to support this outlook too; but just about as many, some of them even more convincing, can be cited for the contention that partnerships make for long and easy survival, a life of continual annuities. The progenitors of some of our most successful modern plants had struck up partnerships with fungi as far back as the Carboniferous age, several hundred million years ago, and the partnership is still going strong. Some of the independent plants and animals that were very successful individualists in that age have long since disappeared and have left no descendants. One may well ask, What price independence? Many of our parasites seem almost discouragingly durable, in the long-time, evolutionary sense, being variable enough so that as their hosts change, they change also. Unquestionably many of these partnerships between plants and animals, or between different kinds of plants or different kinds of animals, have been getting more complex as evolution has progressed. Some of them will eventually specialize themselves out of existence, of course, as will also some of the individualists; others may endure to the end of time, long after the individualists have disappeared. The essential thing, as in bridge or marriage, is to pick the right partner.

Ants and Fungi

There are thousands of different kinds of ants, and they live in diverse ways. Anyone who has the romantic idea that in nature, unspoiled by man, all is sweetness and light and gentle love, should consider the ways of the ants. Wheeler's book on social insects is an admirable introduction to the bright virtues and dark horrors of this strange world toward which we may be drifting. Many of the ants are among the most highly developed of the social insects; compared to our own, their societies are extremely ancient, and some of their caste systems make even the most complex of ours look amateurishly silly. Among the ants, the females do all the work, an arrangement favored even by biologists. The males have one function only, that of fertilizing the queens, and once that is done they are through, finished. Even biologists look askance at this drastic limitation of their biological functions. In the long time that ants have been experimenting with different ways of living, they have had ample opportunity to develop some good social traits that we shall probably not hit upon for a few hundred millions of years, and they have also had opportunity to devise ways of getting food and shelter and pleasure that make our present economy seem extremely primitive. They haven't concentrated on the simple virtues by any means. Some have developed cravings and perverted appetites that lead them to death and destruction. They have also evolved forms of slavery, and some go out and capture other ants to do the drudgery while the master race goes on picnics, to find, when they return, that the slaves have taken over. They also are tricked by other insects that behave to them like brothers, join the clan, take the oath, learn the secret handshake, and then devour them. All of which indicates that ants are not simple. The ones we are interested in cultivate various kinds of fungi for food.

There has been a good deal of speculation as to how some of the ants came to cultivate fungi and depend so completely upon a fungus diet. The most reasonable, but not necessarily the most accurate, explanation seems to be as follows. Most of the ants eat only liquids. They take solids into their mouths, but relatively few of them, outside of those that eat fungi, swallow solids. All ants have a pouch in the lower part of the mouth, called the infrabuccal pocket. Such solid food as they take into their mouths goes into this pouch or pocket. Being cleanly beasts, a part of their leisure time is spent in wiping bits of debris off their heads and bodies. This dirt is collected by means of delicate combs on their forelegs, and the combs are cleaned by being carried through the mouth. The particles of dust, dirt, pollen, plant fragments, and other solid matter so gleaned from their bodies, all go into the infrabuccal pocket. So does the solid food they pick up. In this pouch, the soluble materials are dissolved; the liquid solution passes into the esophagus and is swallowed. The screened-out solids that remain eventually fill the pouch, and when it is full the contents are cast out, much as all of the Americans encountered by Dickens would dispose of a spent chew of tobacco. Often this waste material is disposed of in a special part of the nest, a sort of Viking kitchen midden.

While in this pocket, the solid material is moist and warm. Considering the variety of materials that many ants sample, it is not strange that within the pocket there should be spores and mycelium of a good many fungi. Some of these fungi cannot survive there, others just survive, still others grow. When these waste pellets are deposited in a nest that is warm and humid, as the nests of many ants are, fungi often grow and cover the material. So whether ants like them or not, many are growing fungi unintentionally. Since certain of the fungi do have a rather high food value, it is not surprising that they should be attractive to ants, and so should be eaten regularly and in quantity. And since these edible fungi can be grown within ant nests, and the ants can partially or completely subsist on them, it is not at all to be wondered at that some of the smarter ants should have learned to cultivate them. After all, it took us only a few thousands of years to learn to cultivate some of our present food plants; the ants have had more time than we have, and most of it they have spent working, not talking. We cannot yet compete with them in cultivating fungi for food. In their having taken the initiative and hit upon such cultivation, there really is nothing so very mysterious. But one may well wonder at the expert methods that some of them have devised.

There are about one hundred kinds or species of leaf-cutting ants known, mostly in the Americas. The majority of them are tropical and subtropical, but a few occur as far north as New Jersey and New York; so they are not at all exotic, nor any more fabulous than many other things that occur in those regions. If they are relatively little known to the public at large, it is mainly because they are so quiet. Not all of these leaf-cutting ants are cast in the same mold, of course, but all share in common the practice of cultivating fungi for food. Apparently it is a rather ancient practice.

Each kind of leaf-cutting ant requires, or prefers, the leaves of only certain particular varieties of trees or shrubs, although there may be considerable leeway in this. Presumably their preference depends mainly on what sort of compost their particular cultivated fungus thrives on. Some of what are ostensibly leaf-cutting ants cut no leaves whatever, but go out and collect the excrement of caterpillars that have eaten leaves, bring this material and stray leaf fragments home, and cultivate their molds on this. Much as we cultivate mushrooms on the excrement of horses. The typical leaf-cutting ants, sometimes called parasol ants, from the fancied resemblance of the leaf fragments they carry to parasols, go out in droves to collect leaves.

They gnaw out portions of leaves as large as they are able or willing to carry, march home with them, and go out for more. Like most ants, they are strongly unionized, and those that cut and carry leaves do nothing else but that. Members of a different union, or caste if you prefer, chew up these leaves and pack them into beds for the fungus to grow on, the fungus that forms the sole food of the entire colony. To enrich the compost on which the food fungus grows, they add the contents of their infrabuccal pockets — composed in this case of just miscellaneous dust and dirt — their own dung, and the carcasses of their brethren. This serves not only to keep the nest tidy and promote the growth of the fungus, but to conserve nitrogen. Nitrogen, one of the vital elements in all protein, is as necessary to the formation of ants as it is of human beings. In some regions where these ants live, nitrogen is in relatively short supply, and they are conserving it as we, by processing sewage, have only recently, after a fashion, learned to do, or as some of the older civilizations long have done by using night soil to fertilize fields and gardens and so increase their food supply and their diseases. The ants ages ago learned the virtue of economy in some of the essential elements, and apparently are little troubled by intestinal ailments.

These ants do not cultivate just any fungus, hit or miss, but certain specific ones. Closely related species of ants may cultivate the same kind or closely related kinds of fungus, while more distant relatives cultivate different ones. On the leaves brought into the nest for compost there must be hundreds of different varieties of fungus, as well as a heavy population of bacteria and other microscopic forms of life. These nests are made in moist soil, where fungi abound. One would think that occasionally some wild or weed fungi would take over and ruin the ant gardens, as weeds ruin our own gardens and fields. Perhaps crop failures do plague these fungus-growing ants, but if so we do not yet know of them; the ants seem to be able to keep pure cultures of certain fungi growing in their gardens for years or decades together. Even with all the facilities available in our modern laboratories, some finesse is required to maintain pure cultures of fungi and, in the growing of cultivated mushrooms, weed fungi occasionally overrun the beds to the detriment of the mushroom crop. So it appears that these ants are not altogether stupid in being able to maintain for decades on end essentially pure cultures of their food crop.

Their fungus gardens vary greatly in size. Some of the fungus-growing ants in the southwestern United States and northern Mexico make chambers from a half-inch to ten inches or so in diameter, from the roofs of which they suspend hanging fungus gardens, like draped curtains. The soil of these gardens is made up mostly of caterpillar excrement and small bits of leaves picked up on the ground. Some of them make a series of such hanging gardens, in chambers one below the other in the soil, so that as the soil dries out from above in summer they can retreat to the lower gardens and still have ample food. In the tropics, where moisture is available and leaves for compost are at hand all the year around, some of these ant gardens are larger than many of our homes. One of them by actual measure was a bit more than 300 cubic yards in volume, or about 30 by 30 by 10 feet, a sizable nest indeed. In such a colony there were more than half a million ants, all of them nourished solely by their cultivated fungus. The fungus was and is their only source of food. Special workers of a particular caste or union tend these gardens, and under their care the fungus continually produces an abundance of rather peculiar, pear-shaped cells that are rich in food. We have not yet tried to grow these fungi to find out what we could get from them in the way of special foods or vitamins, but it might be profitable to do so. Recent evidence indicates that some of the fungi cultivated by termites have vitamins not found elsewhere, and these may eventually find some use in our own economy.

When a young queen from one of these nests goes out to found a new colony, she has her infrabuccal pouch filled with the fungus on which the colony has lived for millions of years. After being fertilized, she seeks out a likely spot, grubs out a small chamber, and casts her wad of fungus out on the floor. As soon as the mold begins to grow from this pellet, she lays an egg or two, crushes these, and mixes them with the growing fungus. The eggs are rich in nitrogenous food, and they naturally improve the compost on which the fungus is growing. She also mixes her excrement with the compost, and so gives it still more rich nitrogenous food. This is not done just casually and by chance, but, so far as can be gathered by watching the business, deliberately and by intent. Once she has the miniature fungus garden growing to her satisfaction, the queen concentrates on egg laying. The eggs hatch into larvae, which eventually become ants, and these go out and bring in more food to be made into compost on which the fungus is grown.

There can be no doubt that this partnership between ant and fungus enjoys a certain amount of biological success. Compared with some of the independent ants, however, these fungus-cultivating ones are about like some of the orchids compared with ragweed. Both the orchids and the leaf-cutting ants have formed close partnerships with fungi, partnerships that make for intensive and local, rather than extensive and general success. They have become specialists. As time goes on, they may become more and more specialized. Like Monotropa they have given up freedom of choice for security, a security that is closely akin to slavery, and as with other forms of slavery, the master is the one most in danger of elimination. We must remember that such associations were not arrived at in one sudden revolutionary jump, but rather by a series of infinitely small adjustments. We do not know what the circumstances were, but it is easily possible that the only freedom of choice the ants had was to go into partnership with fungi and risk eventual extinction, or fold up right then; either security would be attained at the sacrifice of independence, or the species would perish. No one can foresee what conditions will prevail a million years in the future, but it is a safe bet that some of these successful partnerships between fungi and insects still will be doing business, and that many new partnerships will have been formed.

The independent and individualistic plants and animals may eventually become a dwindling minority, if they are not already. Ever since the first primitive family was founded, we have been gradually losing certain independences, gaining others. Most of us now would be almost completely helpless if we were deprived of the services and goods of others. For security we have given up freedom, even as these ants. Often the choices have not been deliberate ones, but have been forced upon us by circumstances over which the individual has had little control. Some of the Australian bushmen have much more individual freedom of certain kinds than we do, but even Thoreau was willing to go no farther away from civilization than Walden Pond. Should conditions for the leaf-cutting ants change so that the fungi on which they feed cannot survive, not only the ants themselves, but all of those who have cast their lot in with them — parasites, predators, associates, friends, and dependents — will pass quickly and quietly into limbo. Neither faith, reason, nor experimental evidence may enable us to avoid similar pitfalls.

Some of the termites (whom the ants hate) also cultivate fungi for food. Termites are not ants, but are fairly closely related to cockroaches, and, like the cockroaches, they are among the most primitive insects. Our common cockroaches are independent, rugged individualists that thrive throughout the civilized world.

Some of their termite relatives thrive mightily too, but mainly by virtue of a number of partners. They depend on protozoa in their intestines to digest the wood they eat, much as cows depend on bacteria in their rumen to digest the cellulose they eat. Those termites that in the tropics and subtropics consume wood and paper of all kinds, do better if the wood they eat is rotted by fungi. From the fungi apparently they get both vitamins and partially digested food.

In the tropics some of the termites make tremendous nests, in portions of which they cultivate mushrooms or other fungi for food. Unlike the leaf-cutting ants, they do not depend for food solely upon the fungus, but often they cultivate it in much the same manner, though on not so large a scale, in special chambers within the nest. That is, certain rooms within their nest will be devoted to the growing of fungi, and a special caste of workers will care for these gardens. One kind of termite builds aboveground nests of adobe clay that are shaped almost exactly like a giant mushroom — a rather remarkable case of nature imitating nature. Usually the fungus is fed to the young of all castes, but after the young have reached a certain stage they get no more of this invigorating food. The royalty, or reproductive castes, are fed upon it continuously. Not only is it the privilege of the termite upper crust to eat wantonly of mushrooms, but it is this diet that in part makes them socially elite. It recently has been found that a fungus cultivated by one kind of termite contains special growth-promoting vitamins. Obviously these fungi on which ants and termites subsist should be investigated more thoroughly: what is of such great use to them may be of some use to us, not only in the way of commercial gain, but also in advancing the good life. A very moderate subsidy in cash, and a larger one in imagination, would suffice to explore the possibilities in this wide-open and fascinating field.

Ambrosia Beetles

Ambrosia beetles inhabit recently dead trees, freshly cut logs, newly sawn lumber, and occasionally the stems and branches of living trees and shrubs. The female ambrosia beetle usually has the job of preparing the home, which is simply a tunnel or gallery excavated in the wood. The gallery may be communal, the eggs being laid in groups in large numbers in a single-vaulted chamber, where the young grow up *en famille*, or it may be in the shape of a tunnel in which each egg is laid in a separate niche cut in the tunnel wall. Regardless of the type of dwelling they construct, none of the ambrosia beetles eat anything but the spores of a fungus they cultivate. They were given the name of ambrosia beetles because when they were first studied, about a hundred years ago, an acute student recognized that they fed upon material that seemed to form of itself on the walls of their tunnels. This investigator, for all his acumen, was innocent of any knowledge of fungi, a not surprising state of affairs about 1840. The doctrine of spontaneous generation was still in vogue among many scientists then, and since the material on which these beetles fed seemed to have no earthly source, and as ambrosia was the food of the classic gods, the stuff was called ambrosia, and the beetles which lived on it ambrosia beetles. We do not know whether the classic gods lived solely on fungus ambrosia and fungus-fermented wine, but we do know that fungi are the sole source of food of all ambrosia beetles.

If the beetle excavates a communal chamber, the fungus is planted on the walls of this chamber. It grows a short distance into the wood and usually stains the wood a dark color. On the wall of the chamber the fungus forms multitudes of pear-shaped spores, much like the food cells formed by the fungi of the leaf-cutting ants. After the young insects hatch, they feed on these spores, and the more they eat, the more there are formed. To assure themselves of an adequate food supply, all they have to do is eat. A beautiful arrangement. The fungus associated with the ambrosia beetle common in the twigs of the tea bush has been cultivated, and it has been found that if the surface of such a culture is scraped, the fungus quickly produces large numbers of spores. The beetles associated with this fungus were raised on such cultures, with no other food than they got from the fungus spores, and they lived a full and complete life. These beetles require proteins, minerals, carbohydrates, vitamins, and a few minor and more obscure materials, even as we do, and they get them all from this one cultivated mold that grows on no more nourishing a substance than wood.

Not all of the eggs laid by a female ambrosia beetle in a communal chamber hatch at the same time, and so the young and old ambrosia beetles may be scurrying around amongst the unhatched eggs and newly hatched larvae of their family. They avoid bumping into or injuring the delicate eggs and larvae, and are even ready to defend the helpless young from injury by interlopers or marauders. It is difficult to imagine a beetle, even one smart enough to cultivate fungi, with leisure and an inclination to cultivate social graces, but these beetles seem to have a rollicking and happy family life and to be genuinely attached to one another. Perhaps the fungus food is responsible for this charming state of affairs. Certainly if something in the food or drug line could be found that would make men agreeable without making them drunk, irresponsible, amatory, or immoral, something that would give them a genuinely uplifting stimulus with no hangover the next day and no other ill effects on body, mind, or spirit, there would be a great demand for it. No one has seriously hunted for such an elixir, and one good place to hunt might be among these ambrosia fungi.

If the mother ambrosia beetle lays eggs in individual niches gnawed in the tunnel wall, the food is furnished in another way. Each egg is put into its own niche; then the opening is packed with wood chips, excrement, and fungus. The fungus grows, and when the egg hatches, the young larva finds a crop of nourishing fungus cells ready to eat. When this is all eaten, the spent pellet is removed by the solicitous mother and replaced with a fresh one, so that the infant may at all times have sufficient food. If the larva makes its own tunnel, as some of them do, the fungus develops on the walls of this tunnel, so that all the larva has to do is eat and grow. We have grown the fungus associated with an ambrosia beetle that inhabits one of our common forest trees, and have found it difficult to cultivate in the laboratory, even in wood with holes bored in it to simulate the tunnels made by the insects. Apparently this fungus is almost as dependent on the beetle as the beetle is on the fungus.

So long as the gallery or tunnel system is occupied by actively feeding beetles, their food fungus thrives, and no other fungi compete with it to any extent. Since the tunnel is open to the air at one end, the spores of many common air-borne fungi must enter it at one time or another. The beetles themselves also probably carry in many different kinds of organisms, including a variety of fungi, when the tunnel is constructed. Yet none of these weed fungi get a foothold so long as the tunnel system is occupied by young, growing, hungry ambrosia beetles. How they maintain such pure growths of their special cultivated fungi is still their own secret. A reasonable guess would be that relatively few kinds of fungi can grow very well in that special environment (most ambrosia beetles are very fussy about the water content of the wood in which they make their homes) and that of those fungi which do grow, only the food fungi can stand continuous cropping by the beetles. There may be more subtle things involved, for in biology the most simple solution seldom is the correct one. Once the beetles have gone, weed fungi are likely to fill the tunnel, unless the wood becomes so dry that none can grow there. Before leaving the ambrosia beetles and their food fungi, it should be mentioned that some of them have on their heads structures like miniature baskets in which they carry spores of their special fungus from their old home to a new one.

A number of beetles related to the ambrosia beetles also carry fungi around with them and eat these fungi to some extent, but mostly depend on fungi to kill trees for them and thus give them places in which to raise their broods. A number of bark beetles common in coniferous or evergreen trees bore through the bark of living trees, inoculate the wood with molds which stain the wood blue and kill the trees, and so provide homes for more bark beetles. These beetles generally do not thrive in living trees, but if they can penetrate a living tree far enough to inoculate the sapwood with the spores they carry, the fungus will kill the tree, and thus permit the beetle to increase tremendously.

Dutch elm disease is carried mainly by a beetle of this type that has found a fungus useful to it. This beetle does not invade the bark of a healthy tree, but uses a more subtle and indirect attack. A newly emerged adult, covered with spores of the fungus that causes Dutch elm disease, will feed on the young and tender buds of a *healthy* tree. What it gets from such feeding we do not know — probably vitamins of one sort or another. As it bites into the bud, or into the twig just below the bud, it inoculates the tree with the spores on its whiskers. Then it goes off to raise a family under the bark of a dying or recently dead tree. The fungus inoculated into the tree by the beetle grows down into the twig, then into a branch, then into the trunk. In a few years it kills the tree, which is then invaded by the grandchildren or great-grandchildren of the beetle that inoculated it by feeding on the bud. The fungus forms masses of sticky spores all over the walls of the pupal chambers in which the beetle is transformed from a larva into an adult. The young adult, to get out into the open air, has to gnaw through this layer of spores of the Dutch elm disease fungus, and so when he emerges into the world, his whiskered head and body are covered with spores. These he carries to a healthy tree in the neighborhood, inoculates the tree by feeding on the buds, and so prepares a home and breeding place, not for himself and wife, but for the members of a generation several years in the future. Many of our plant disease problems are complicated by associations between insects and molds; those interested in following up some of the technical aspects of this relationship should consult the excellent books by Büchner and Leach. See also Plate 5 facing page 120.

Round Trip of a Molecule of Nitrogen

So far we have seen that a tree which lives by virtue of a fungus on its roots may support on its twigs and branches a population of scale insects that also depend on a fungus for housing and protection from enemies. Its leaves may be stripped by leaf-cutting ants as compost for their fungus food. It may be killed by fungi brought into it by insects. Eventually it will be rotted by fungi that convert the wood again into the soil from which it came, soil in which fungi are among the dominant flora, consuming other fungi, bacteria, protozoa, insects, and being in turn consumed by them.

No one has yet followed a marked molecule through its cycle, and so the following must be a largely hypothetical, if perfectly plausible, account. A molecule of nitrogen, as a nitrate salt in the soil, is grabbed up by a fungus, is converted into protoplasm, and flows through the narrow, branched cells of fungus mycelium into a mycorrhizal root. There it is transferred to the tree, goes up with the sap stream, and ends up in a particular cell in a leaf. The leaf is cut by a leaf-cutting ant, chewed up into compost, digested by the cultivated mold, and transferred to a food cell, which is eaten by, and goes into the construction of, another member of the ant colony. This ant, out on a foraging trip, is eaten by a bird, and our N molecule, having become part of the bird, eventually goes into an egg. The egg is eaten by a weasel, who comes to a bad end, and his carcass is consumed by carrion beetles, the N molecule that we are following going into a beetle. The beetle dies of undetermined causes and is invaded by a fungus, into which the N molecule goes; the molecule ends up in a spore of this mold, which spore is eaten by a nematode and digested, the N molecule being built into his body. The nematode is trapped by a predatory fungus in the soil, and by chance our molecule of N is released as part of a molecule of ammonia. This is converted into nitrate and picked up by the roots of a green plant, and the cycle starts over again.

That would be a fairly simple round trip. Actually the N molecule might cross continents, pass through hundreds of different kinds of organisms, and explore most of the seven seas on its way from soil to soil. But at one time or another it would be likely to pass through one or more fungi. Most living things are built up of the same elements; all life needs such things as nitrogen, potassium, phosphorus, iron, copper, zinc, sulphur, and so on. Some of these elements are available in only relatively small amounts, and for these there is a continual battle. Only by their successive transfer from one organism to another can the whole biological structure survive. Thus what to us are death, decay, and dissolution are, to those organisms causing them, a time of plenty and rejoicing, of banquets and bounty. We decay and rot the food we eat just as the molds decay and rot the food they eat. One of our functions is to keep the business of decay balanced as heavily in our favor as we can, and to see that we rot as much material as some of our more vigorous competitors, including the fungi. One of the reasons that things digestive are not exactly socially acceptable topics may be that we instinctively recognize that our nourishment is essentially a business of rot and decay, a destruction of other forms of life. The N molecule even gets involved in religion when it gets into animal protein on its round trip.

Pilobolus, the Cap Thrower

It is stretching a point to bring this fungus into a discussion on fungus partners of animals, for while it undoubtedly gets great benefit from the animals it inhabits, there is no evidence that the animals concerned are more than just passive carriers of the fungus, environments in which the fungus can thrive. But the fungus is in itself so interesting that it would be a shame not to tell about it, and this is as good a place as any. The name *Pilobolus* means *cap thrower*, and it is a descriptive name. Relatively few people, aside from mycologists and a few of their students, have had the fun of watching the minor miracle of Pilobolus. Anyone who is not too squeamish to handle horse or cow manure can see the whole amazing process for himself and know the delight of learning at first hand what goes on in one part of nature. All you need is a modicum of curiosity, a stone jar or bucket of convenient size, and a nearby horse. The breed, social standing, age, speed, location, or contours of the horse do not matter; but the dung must be fresh, preferably only a few hours old. To raise a crop of Pilobolus, proceed as follows: Collect a quart or two of fresh horse dung, put it in the bottom of a gallon jar or pail, and cover this with a piece of glass. Put a match stick or bit of paper under one side of the glass lid, to allow sufficient circulation of air so that Pilobolus will not be smothered. Place the jar in a warm room, preferably where the temperature is around 80 to 85 degrees Fahrenheit; if the room temperature is in the low 70's, a desk light a foot or two above the jar, with the reflector adjusted to cast the light into it, will be good enough. If the air is dry, sprinkle the dung lightly with water when it is put in the jar, and again twenty-four hours later.

Between the second and fifth days, depending on the temperature, you will find, when you look through the glass lid around 8:30 in the morning, that the dung is covered with hundreds of delicate stalks, each of them less than a half millimeter in diameter and from one to two centimeters long. All of them point upward, directly at the light. In technical jargon, they are phototropic, which means that they grow toward the light. That in itself is not unusual — most plants do, including a fair number of fungi. If you look closely, you will see that each stalk is surmounted by a swollen tip, and on top of this is a hemispherical, black cap, which in a couple of hours is going to be shot into the air.

Although many plants grow toward the light, few of them aim at it so directly and precisely as Pilobolus. The swollen cell beneath the black cap acts as a lens. Just at the base of this lens is an "eye," consisting of a bit or orange-colored pigment, and this regulates the direction in which the stalk grows. The stalk grows in length just below this light-sensitive eye. As the stalk grows upward, the tip points consistently toward the source of brightest light. You can prove this by shifting the light from side to side during the early morning hours when the stalk is elongating. The stalk will grow in a zigzag fashion. Considering that this fungus is supposedly at the lower end of living things, it has some effective and subtle adaptations, and this eye is only one of them. It is superior to the eyes of some of the lower animals. The utility of this to Pilobolus is obvious. In nature, horse dung often is deposited in pastures where large-leaved weeds are common. The eye described above insures that Pilobolus will not shoot its spores against an overhanging mullein leaf, but out into the open air at one side or the other of the leaf.

By about 9:30 in the morning the hundred thousand or so spores in the black cap of each stalk are mature, having been formed within a few hours, and the fungus cannon, or mortar, is ready to shoot. In preparation for this, several things go on simultaneously to make the shooting effective. A groove is formed in the wall of the swollen portion, just below the cap, and this makes for a regular, uniform, and simultaneous break in the wall all around the circumference of the cap. The swollen cell exudes droplets of water, the salts in the sap within it thereby being concentrated and the pressure increased. Eventually a pressure of several atmospheres is built up. Under this tension the wall breaks along the groove, the swollen cell explodes with an almost audible "Plop!" and the cap is sent flying through the air. Plate 4 and Figure 5 will help you to visualize this process.

The force of this miniature burst is great enough to send the cap off at a speed close to thirty miles an hour. Most of the caps are shot off between 9:30 and 10:30 in the morning (you can delay this until afternoon by putting the jar in which the fungus is growing into the refrigerator when the guns are about ready to shoot). At this time of day, during the time of year when the fungus is likely to be active out of doors, the stalks, pointing at the sun, have an angle of about 45 degrees. As any artillery student knows, this gives pretty close to the maximum range for a projectile — a fact which the simple Pilobolus long ago discovered.

On an object as small as this cap — just slightly larger than a fly speck — the friction of the air is high, and so the original, or muzzle, velocity of the flying cap is quickly reduced. Even so, these tiny capsules of spores are shot an average horizontal distance of from four to five feet, and we have measured an occasional shot of from eight and one-half feet to nine. As in other

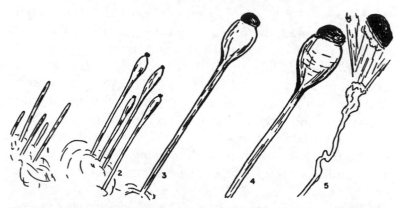

Figure 5. The course of development of the "gun" of Pilobolus.
(1) Single stalks, the sporangiophores, grow upward. (2) The
tips of the sporangiophores begin to enlarge, and black spore
cases, the sporangia, form on top. (3) A nearly mature gun.
(4) The gun is mature and ready to shoot off the sporangium
into the air. (5) The gun at the moment of shooting. The black
sporangium has been discharged into the air, the stalk collapsing
with the recoil.

shooting, accuracy as well as distance enters into the final result,
and in the accuracy test Pilobolus does moderately well. This
can be determined by a simple experiment. The jar in which the
dung is placed may be surrounded by a cylinder of stiff black
paper a couple of feet high, or placed in the bottom of a fifteen-
gallon barrel or drum, where light can enter only from the top.
The top of this should be covered with a piece of white paper,
on which the black spore-caps will show up readily. Over this
a piece of heavy black paper should be placed, with a hole two
inches in diameter cut out of the center. Thus the only light
reaching Pilobolus will come through this bull's-eye. The glass
cover, of course, has to be removed from the jar the night before
the fungus is to shoot; else the light will be diffused by the glass
and impair the mold's aim.

After the fungus the next morning has finished shooting, the
white paper, with the black projectiles glued to it, can be examined. Often 50 per cent of them will be within a two-inch bull's

eye a couple of feet away, a vertical distance that is roughly 75 times greater than the length of the fungus stalks. The actual shooting part of the fungus, though, is just the swollen cell, and this is from one fifth to one tenth of the total length of the stalk; so a vertical range of two feet is from 375 to 750 times the length of this portion. Also the stalk is slender, thin-walled, and anchored at its base by nothing but a few delicate fungus threads. The recoil from the burst is strong enough to blast the stalks flat against the substrate from which they are growing. All in all, this is not bad shooting — many a deer and duck hunter brags all winter on less.

There are still other refinements. The spore capsule is approximately hemispherical in shape, rounded on the upper side, flat below. As it is shot off, a bit of very gelatinous material adheres to its base. A drop of sap from the cell that exploded also clings to it. No matter how the cap lands, it turns so that the flat side lies against the surface on which it has hit. The droplet of cell sap that adheres to the flat side is relatively heavy, and so if the rounded side strikes the landing surface first, the weight of the droplet pulls it around. Also the rounded surface of the cap is nonwettable, and so even if it strikes a leaf with its apex, the droplet of sap will splash down on the leaf, and pull the cap around. As this sap has a wetting agent in it, it quickly spreads out in a thin film on even a waxy grass blade, and dries almost instantaneously. The bit of gelatinous glue on the under side of the cap then sticks the cap securely to the surface on which it has landed — so securely that after it has dried for a few moments it is not possible to wash it off even if it is left under a running faucet for a couple of days.

Thus in nature these caps, stuck to the leaves and stems of a large number of plants from three to eight feet east by southeast (in the Northern Hemisphere) of the dung on which they were borne, remain on the plants for a long time. Eventually some of these plants will be eaten by a horse. What happens to it then, no one yet knows. We have never failed to get Pilobolus from the dung of horses on the University Farm in St. Paul, horses which almost never see a green pasture, and which are fed on hay from fields where only tractors gambol. Even circus elephants have their own kind of Pilobolus on their dung.

Much of what is known about Pilobolus has been learned from the work of Professor A. H. Reginald Buller, of Winnipeg, Canada, who for years was one of the world's leading mycologists. He spent his time not in putting dead and dried specimens in pigeonholes, but in prying into the secrets of living, growing fungi, and he brought them to life as few other people have. Nearly the first half of Volume 6 of his *Researches on Fungi* is devoted to the Pilobolus gun. Were there as many students of fungi as students of the mathematical sciences on the national and international committees that award prizes for original research, Dr. Buller would have been loaded down with medals.

Pilobolus is not the only fungus that grows on dung. It will bear a heavy crop of its cannons each morning for three or four days, then give way to others. Some of these have also been studied, by Dr. Buller and others, but not all of them are known thoroughly. In a recent study at one of our older southern universities, more than sixty different kinds of fungi were found on the dung of various animals. Some of these probably have adaptations and devices as marvelous as those of Pilobolus, waiting only for some future student with the genius and devotion to discover their ways of life. To many this prying into the secrets of the minor marvels of nature appears futile, a business of learning more and more about less and less, but this yearning to find out what makes nature work, and why and how, is the basis of all research. If we are eventually to control nature, we must know something about how nature operates. An intelligent and thorough study of the most humble dung fungus will add something to our accumulated body of knowledge, our understanding, our culture, even as does the study of minor stars whose light started off to us a couple of hundred million light years ago, or the forces in the atom, or what makes music. Sometimes the facts are of value in themselves, sometimes they are interesting, now and then someone finds a key that opens up a whole new field. The disciplines involved in studying a dung fungus intelligently are just as severe and just as civilizing as those involved in some of the more practical lines of research. Science is mainly the experimental approach, and before science can remake the world in any beneficial sense the objective and critical attitude will have to become more widely diffused than it now is. One way of acquiring it is to study something thoroughly, even a dung fungus; to learn everything that your intelligence and ingenuity and time and facilities will permit you to learn about it; and learn it so well that someone else can duplicate your tests and get the same results. This engenders a wholesome skepticism about many things, including your own opinions and beliefs — a skepticism that need not be, as some have suggested it is, accompanied by spiritual withering.

Additional Reading

Paul Büchner, *Tier und Pflanze in Symbiose*. Berlin: Gebrüder Borntraeger, 1930. 900 pp.

A. H. Reginald Buller, *Researches on Fungi*, Vol. 6. London: Longmans, Green & Company, 1934. 513 pp.

John N. Couch, *The Genus Septobasidium*. Chapel Hill, N.C.: University of North Carolina Press, 1938. 480 pp.

J. G. Leach, *Insect Transmission of Plant Diseases*. New York: McGraw-Hill Book Company, 1940. 615 pp.

Alfred Möller, *Die Pilzgärten einiger südamerikanischen Ameisen.* Botanische Mittheilungen aus den Tropen, Heft 6. Jena: Gustav Fischer, 1893. 124 pp.

CHAPTER 5

Fungus Parasites of Plants

As we have seen in the earlier portion of this book, fungi live upon many sorts of things, and get their food in many different ways. A considerable number of them live upon and get their food from living plants. They parasitize other plants, and cause diseases of them.

Roughly, these fungus parasites can be divided into "facultative" parasites and "obligate" parasites. The facultative parasites are those fungi that live partly on dead materials, mostly plant remains, and partly on living plants. They can live a full and complete life on the debris in the soil or in decaying vegetation in the water or in the dead trunks or roots of trees. They do not require living plants on which to subsist. But if they encounter susceptible host plants, or if normally resistant host plants are made susceptible by unfavorable growing conditions, or if conditions result in a heavy concentration of them, they may quit their saprophytic mode of life and invade living plants, thereby becoming parasitic.

The majority of fungi that cause disease in plants are "facultative" parasites, and some of our most destructive plant diseases are caused by them. Many of them are difficult to live with, yet almost impossible to eliminate. A number of them, for example, that live on plant debris in the soil, attack and kill the roots of almost any kinds of plants that we grow. Because of the variety of the plant debris on which they can live, they are widely distributed throughout the world. They have a wide range of appetite, both as saprophytes and as parasites. Normally inhabiting dead material in the soil but not averse to a diet of living roots of plants when conditions favor it, they have caused a heavy reduction of the yield, and also crop failures, in such diverse crops as forest tree seedlings, sugar beets, peas, alfalfa, brome grass, flax, muskmelons, bananas, tomatoes, those grasses used for lawns and golf greens, and various sorts of flowers, to name a few. The list could be extended indefinitely. Perhaps a better idea of the range of attack of the soil-inhabiting fungus facultative parasites could be given by saying that we do not know of any crop plant that is not subject to more or less regular attack, and occasional killing, by such fungi.

Not all of the facultative parasites among the fungi normally live in the soil. Many of them attack the leaves, stems, flowers, or seeds of various kinds of plants, or invade the bark or the living portion of the trunks of trees. These fungi are common, not rare. One can find them almost everywhere upon the living portions of almost all kinds of plants, all through the growing season. Invasion by these facultatively parasitic fungi is one of the natural hazards to which all wild and cultivated plants are exposed. It is part of the "dog eat dog" plan of nature. It may be neither desirable nor beautiful, but it often is interesting.

Some of the fungi are "obligate" parasites. That is, they grow only in the living tissue of the living host plants, and nowhere else. Most of these obligate parasites among the fungi are highly specialized as to the host on or within which they can grow. This will be enlarged upon later in the chapter, but as an illustration of the principle, the rust fungus that attacks cereal crops is restricted to cereal crops, and will not spread to flowers, shrubs, or trees. Mildew of phlox will not spread to other flowers in the garden (except possibly to some very closely related to phlox), and downy mildew of grapes will not attack anything but grapes. Most of the obligate parasites among the fungi are included in the downy mildews, the powdery mildews, and the

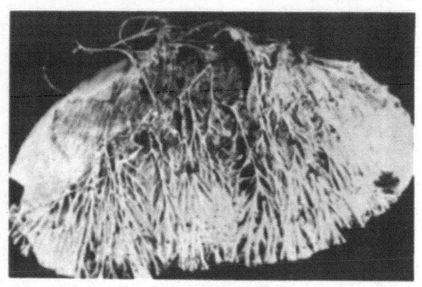

Plate 1. Mycelium of a wood-rotting fungus. The mycelium grew in through an almost invisible crack in the concrete wall of a basement and then invaded wood and other materials. The mass of mycelium here photographed was growing over a wire screen cage, and was more than a foot across.

Plate 2. Spores of Alternaria, arising from branches of the mycelium.

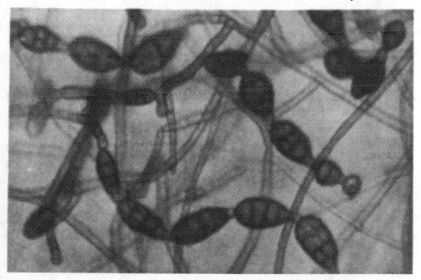

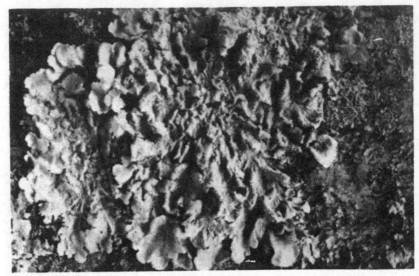

Plate 3. A common lichen (*Parmelia*) growing on the bark
of an aspen tree.

Plate 4. A mass of fruiting stalks of Pilobolus about to shoot off their
sporangia. The photograph was taken at about 9:15 A.M.; less than half
an hour later all the sporangia had been discharged.

rusts. Their rather extreme specialization as to the host on which they will grow would seem to limit and restrict them to a relatively minor status as plant enemies. And before man began to cultivate various kinds of crop plants intensively, they may have been, and probably were, minor enemies of them. Probably they were relatively unimportant factors in the establishment of natural plant societies, although this is not certain. But in our cultivation of many plants they have now certainly become major factors. Perversely enough, their extreme specialization, which would seem to limit their increase, is just another facet of their extreme variability, which enables them to just about keep up to our progress in breeding varieties of plants resistant to them. They are extremely shifty enemies.

The varied nature and cussedness of both facultative and obligate parasites is best made clear by individual examples, to which we shall now turn.

Fungus Parasites of Fungi

The old but biologically sound jingle that "Big fleas have lesser fleas upon their backs to bite 'em" applies just as much to fungi as to fleas. Some fungi prey upon other fungi. Fungus parasites of fungi have never merited special study, and therefore probably only a few cases of this sort of parasitism, out of the multitude that must occur, have been described. They have been encountered more or less by chance, or because they were very obvious. A few of these will be described.

The fungus *Dispira cornuta* was, until relatively recently, thought to be a rather rare plant. Before about 1930 it had been found once in Germany, and some years later was found in America, growing upon a not too distantly related Phycomycete. An inquiring mycologist at Harvard encountered this Dispira growing upon Mucor that was growing upon animal dung, and began to investigate it. He found it to be a common and widespread fungus, a regular parasite of Mucor and some other Phycomycetes isolated from the dung of pigs, dogs, and other domestic animals. It was, like many another "rare" plant, common if one knew how, when, and where to look for it.

Rhizoctonia is a genus of fungus that inhabits the soil through-out the world. Some strains of it invade and kill the young roots of many kinds of cultivated plants. Recent evidence proves that they also will invade and kill certain other kinds of fungi. The mycelium of Rhizoctonia penetrates through the wall of the my-celium cells of other fungi, grows in the interior of these cells, and kills them. There is a bit more here than meets the unaided eye. For decades we plant pathologists have been in the habit of considering Rhizoctonia as a rather undesirable member of the soil microflora. We knew that some strains or varieties of it were harmless to crop plants. We also knew that many strains or varie-ties of it were very destructive in that they would invade and kill the roots of crop plants, from peas to pine trees. Now it seems that this same fungus will invade and kill other fungi in the soil. Some of these fungi that it invades and kills are those that attack plant roots. If we inoculate soil with this particular strain of Rhizoctonia we may eliminate those fungi that would otherwise kill seedlings or kill and rot the roots of mature plants. So far this has had no practical application: it is so new that it has hardly got into scientific literature, much less into actual agricultural practice. It may not turn out to be of much practical importance, and then again it might. There is the possibility that we could inoculate soil with this particular strain of Rhizoctonia and thereby eliminate certain troublesome root rots of plants that are grown intensively in greenhouses and fields. Still more important is the idea or principle got from it that we might use a beneficial fungus to control detrimental ones. This aspect of soil micro-biology is still relatively new, but it may be important for the future, and it will warrant thorough exploration.

This is only one example. Another interesting one is offered by certain strains of a fungus known as Chaetomium. Chaetom-ium also is a common fungus of the soil, where it lives mostly upon plant debris — bits of cellulose in the shape of fragments of stems and leaves and wood. It is one of the most avid consumers of cellulose among the microorganisms, and it often gets out of the soil and invades the cellulose fibers of fabrics. Recently a strain of it was picked up from seeds by an enquiring student, a strain that proved to be very antibiotic to many other kinds of fungi. It apparently did not actually invade these other fungi, but rather produced chemical products that diffused out around it and kept the other fungi from growing, or even killed them. When soil was inoculated with *Helminthosporium victoriae*, a fungus that invades and kills oat plants, and then the same soil was inoculated with this strain of Chaetomium, and oat seeds were planted in it, an almost perfect stand of vigorous plants resulted. When the soil was inoculated with *Helminthosporium victoriae* alone, and oat seeds were planted in it, most of the seedlings were killed, and most of those which survived were diseased. In other words, inoculation of the soil with Chaetomium controlled this destructive Helminthosporium disease — at least, in small experiments in the greenhouse.

Will it do the same thing in the farmer's field? Will other fungi, not yet discovered, prove effective in controlling some of our destructive plant diseases caused by fungi? We do not yet know. But some fascinating and important possibilities have been opened up. This beneficent biological warfare is an entirely new approach to the control of plant diseases, and the future may prove it to be something more than just a laboratory phenomenon.

Cicinnobolus is another interesting fungus parasite. It grows in the perithecia, or spore cases, of the powdery mildew. Powdery mildew fungi attack many kinds of wild and cultivated plants, produce masses of powdery spores on the surface of the infected parts during the growing season and, in the fall, produce spore cases in which the overwintering spores are borne. Cicinnobolus grows in these spore cases, consumes the entire contents, and converts it into a mass of Cicinnobolus spores. The powdery mildews are obligate parasites of higher plants — they will grow only in the living cells of living plants — and Cicinnobolus will grow only in the living spore cases of the mildew fungus. It is an obligate parasite of an obligate parasite. Whether it in turn is parasitized by still other fungi, and they by still others, we do not know.

Some of the rust fungi are often parasitized by other fungi. Sometimes these infections are heavy enough to reduce spore production by the rusts. It was once thought that these fungus parasites of rust fungi might be of some value in the control of rusts, but apparently they are not.

Mushrooms, both wild and cultivated, are attacked by a number of different fungi. The cultivated mushroom is susceptible to a number of rather serious diseases caused by fungi — fungi that either invade the beds or invade the mushroom itself and distort, stunt, or discolor it. One wild mushroom, *Clitopilus abortivus*, was given the specific name *abortivus* because the fruit bodies are so commonly misshapen or aborted: they are irregular nubbins up to a couple of inches across, without definite stem, cap, or gills. And they are aborted because they became infected, when very young, by a parasitic fungus. One may find both normal and infected specimens in the same clump, but in our experience the infected and aborted specimens are much more common than the noninfected, normal ones. Other wild mushrooms in the genera Lactarius and Russula are rather commonly parasitized and distorted by a fungus that grows on them and covers them with a layer of mycelium that suppresses the gills and causes the cap to be malformed. The fungus parasite in this case is Hypomyces. Some species of Hypomyces cover the host mushroom with a white layer, some with a pale, pastel green, some with bright reddish-orange. The common, bright orange one is *Hypomyces lactifluorum*, and it is an obligate parasite upon certain species of mushrooms. Even among the mushrooms themselves there is one, Nyctalis by name, that grows parasitically upon other mushrooms, and one of the jelly fungi, *Tremella mycetophila*, grows only as a parasite upon the stems and caps of *Collybia dryophila*, a common forest and woods mushroom.

Thus the parasitism of one fungus by another is not a rare, but rather an everyday, phenomenon. Fungi in all of the major groups are parasitized by other fungi, and the parasites themselves are not restricted to one taxonomic group, but are scattered at random through the whole fungus kingdom. While we do not know of very many cases of fungus parasitism of fungi — the total number described would probably number only a few hundred — the ones we know about have been encountered more or less by chance. It is likely that many more exist than are yet known, and as has proved true in so many other fields of study, from fungi to philosophy (if they can be separated), what appears at first to be merely interesting and intriguing, may turn out to have overtones of general biological and practical significance.

Fungus Parasites of Higher Plants

Fungi grow as parasites on all kinds of plants, from algae, lichens, and mosses to seed plants. Every species of seed plant, both wild and cultivated, has its burden of fungus diseases. In wild plants these fungus diseases are one of the factors that help to maintain the biologic balance. This biologic balance is a fluctuating thing, not static; plant societies rise and fall even as human societies do; they encroach upon one another's territory by all means, both fair and foul. Fungi are one of the factors — but of course not the only one — that influence the speed and direction of this fluctuation. They often help to determine the success or failure of certain plant populations.

In cultivated plants, where large numbers of genetically uniform individuals are grown together, fungus diseases may become a direct and dominant factor in that they determine how much of what kind of crop can be grown in a given area. Where a given crop is grown on the same land year after year, fungus diseases may increase to the point where they make cultivation of that crop difficult or economically impossible. This is one of the basic reasons for crop rotation. Fungus diseases may actually eliminate the growing of certain crops in certain areas, where that crop could be profitably grown if fungus diseases were not present. And in almost all the regions where our major or minor economic plants are grown, fungi have a good deal to do with the quantity and quality of the crop harvested and the quantity and quality of what finally reaches the consumer. To illustrate how these fungus diseases affect our agriculture and our general economy, a few outstanding ones will be described.

Fungi that kill seedlings. Most of those who plant seeds of any kind, whether of flowers, vegetables, cereal plants, trees, or what have you, assume that all they have to do is plant the seeds and water them and eventually they will have seedlings and finally mature plants. It doesn't always happen that way. Seed is not just seed. Seeds of most plants harbor a variety of fungi. The soil in which seeds are planted is likely to be teeming with molds. Some of these molds in the seed and in the soil invade the roots and stems of germinating seeds and succulent seedlings and kill them. In small home gardens it is a minor matter if 50 per cent or more of the seedlings are killed by molds before the seedlings appear above the ground. Most home gardeners plant seed about ten times as thickly as is necessary, and some thinning of them by the fungi merely saves later labor. But truck gardeners, who buy seed at from twenty to a hundred dollars a pound, florists who may pay more than its weight in gold for the seed of certain flower hybrids, foresters who plant large numbers of seeds costing up to twenty dollars a pound, from which they hope the trees of the future will come, farmers who plant hundreds or thousands of acres of seed for which they have paid several dollars a bushel, can ill afford to plant one hundred seeds to get one, ten, or twenty surviving seedlings. Many still do.

A simple experiment is sufficient to bring out this hidden loss from fungi. Place one hundred tomato or pine tree or radish seeds on moist paper toweling in a covered dish. In a few days they will have germinated, and by counting the nongerminated seeds or the germinated ones, you can easily determine the germination percentage. If the seed is fresh and of high quality, from ninety to ninety-five seeds out of a hundred should germinate. Now plant one hundred of the same seeds in black soil in a pot or can, keep them well watered, and count the number of seedlings that come up. An emergence of 75 per cent would be excellent, but it is not uncommon to find only from 25 to 50 per cent of them coming up. The others have been killed and decayed by fungi before they came above the ground. Technically, this is known as pre-emergence damping off. After the seedlings are above the ground, they still are not safe. Their succulent roots and stems may be invaded by fungi in the soil, and seedlings that one day appear healthy and vigorous, the next day may be dead. Within a week only five or ten seedlings may survive where there should have been ninety or ninety-five. This killing of young seedlings by fungi in the seed or in the soil often makes the difference between a successful crop and no crop at all—in corn, wheat, cotton, peanuts, pine trees, tomatoes, sugar beets, and almost any other plant that we grow from seeds. In any commercial growing of plants, this unseen loss may be a vital factor.

Much of this hidden but very real loss can be eliminated by treating either the seed or the soil with a fungicide that protects the germinating seed and young seedling until it is sturdy enough to be on its own. Literally hundreds of different fungicides have been designed and tested as seed protectants, and a dozen or more of them are now widely used. No one fungicide is best for all crops—one may work well on soybeans, another on wheat, another on tomato seeds. Or a given one may protect soybeans from damping off in Minnesota but not in Illinois, or it may be effective in Iowa in 1949, but not in 1950. The first effective seed treatment was devised about a hundred years ago by a Frenchman, but in that case it was to protect seed from infection by stinking smut, not seedling diseases. Most of the work on seed- and seedling-protecting fungicides has been done in the last three decades. Such seed treatment is now generally used in many agricultural crops. It is one of the significant contributions of science to modern agriculture, and to a large extent it was made without much fuss or fanfare. Many men have had a hand in it, including research men at public and private experiment stations and universities, research men in the chemical industries, extension men, and even editors. It has not received much publicity because it just doesn't partake much of the dramatic, but few modern developments have contributed as much as seed treatment to the stabilization of agriculture.

Fungus wilts. Once plants are established, fungi from the soil may invade the roots, grow up into the stem, decay some of the tissues, and in the process secrete materials that are poisonous to the plant. When these toxins accumulate in sufficient quantity, they cause the plant to wilt and die. A number of different fungi cause such wilts in different plants, but one of the most notorious of them is *Fusarium lini,* the cause of flax wilt.

Flax has long been grown in Europe, mainly for the fiber. Even there it was known as a rather uncertain plant. If cultivated on the same land for several years in succession, it began to fail. It was said to exhaust the soil, to use up the food materials in it and impoverish it so that the flax no longer could get what it needed to grow. Once the value of the oil and the seed cake from which the oil had been expressed became appreciated, flax was on the way to becoming an important and valuable crop. As agriculture moved westward in the United States, flax moved westward with it. It did wonderfully well on newly broken prairie sod, but after a few years the soil became flax-sick, and the cultivation of flax had to be abandoned.

Around 1900 Bolley, at the North Dakota Agricultural Experiment Station at Fargo, had the idea that this flax-sickness might be due not to soil exhaustion, which nobody had proved, but to fungus infection of the roots and stems of flax. He soon accumulated evidence that this was so. He then reasoned that one way of fighting this insidious disease was to grow flax in the same ground year after year, until the soil was filled with *Fusarium lini.* Which he did. After several years only a few plants continued to survive in this heavily infested soil. Seed from one of these, again planted in the same soil, produced a good stand of plants. They didn't "acquire" resistance to wilt. Rather, his original population of flax was genetically heterogeneous. A few of the plants possessed inherent resistance to *Fusarium lini,* at least to those strains of the fungus that occurred in his test plots.

From these resistant plants he got out a variety which he named Bison, the tutelary animal of North Dakota Agricultural College. It was widely grown for decades, and has been generally used as a wilt-resistant parent in breeding new varieties of flax. This wilt-resistant Bison was not the only contribution Bolley made from his study. He also laid the ghost of "soil exhaustion" by certain crop plants. Wherever such soil exhaustion has been thoroughly investigated, it has been found to result from a complex of fungi or other organisms in the soil. When a given crop plant is grown year after year on the same land, these organisms increase, until finally the crop plant no longer can survive. There is nothing very mysterious about it.

In addition Bolley also devised the method of using a "disease garden" — of growing the plants in a heavy concentration of the disease-causing fungi — to select individuals resistant to a particular disease. The principle has been widely and profitably used and has had a profound effect on modern agriculture. Relatively few people outside of plant pathologists have ever heard of Bolley; yet he contributed more to the good and abundant life of this and other countries than, say, most of the heroes described by Plutarch in his *Lives* contributed to theirs.

Flax is not the only plant subject to wilt. Similar wilts, caused by closely related species of Fusarium, at one time practically eliminated the profitable growing of watermelons in Iowa, of muskmelons in the Twin City area of Minnesota, and of tomatoes and peppers in many local areas. They have plagued the growers of peas on a large scale for canning, and have chased the banana-growing industry hither and yon in Middle America.

Why not, in all these cases, develop resistant plants as Bolley did? The problem has turned out to be more complex than it at first appeared. As the wilt-resistant Bison flax was grown for a longer and longer time in more and more places, it became evident that at some times and in some places it was susceptible to wilt. Other wilt-resistant varieties bred and selected by later workers were resistant in some areas, susceptible in others. The reason for this seemingly contradictory state of affairs was found to lie in the fungus that causes the disease. *Fusarium lini* is not just a single uniform species. In any flax-wilt nursery there may be several dozen different strains or biotypes of the fungus. One of these may attack certain varieties of flax heavily, other varieties of flax moderately, still others lightly. Another strain of the fungus may attack heavily some varieties of flax that were immune to the first strain. And so on. A flax variety resistant to nearly all of the strains of the fungus present in the research field at St. Paul, Minnesota, may be susceptible to some strains of the fungus present in Ontario or in the fields of Argentina. This is not theoretical; it has been proved by repeated tests.

Of what significance is this? For one thing, we import flax seed from Argentina. *Fusarium lini* is present in the seed and in the fragments of stems and debris imported along with the seed. Seed and associated debris that harbor potentially vicious strains of the fungus are spilled at terminals and along railroad rights of way. Teams of plant breeders, plant pathologists, and biochemists may put in decades of work producing wilt-resistant flax with a high yield of high quality oil. Because there are so many different varieties of the fungus, none of these varieties of flax will remain resistant forever or in all places. The chance introduction of new and vicious strains of the fungus may negate some of their work. For this reason the work of breeding new varieties is a continuous one, a job that must go on without interruption in every region where flax is expected to be a major crop. It would do no harm for legislators to realize that fungi are not much subject to governmental regulations and that they often pose problems the solution of which requires basic and practical work in different fields for a long time.

Late blight of potatoes. Before 1500 the potato was unknown except in a few regions of South America, where it had long been cultivated by the Incas in the Andes. According to statistics of the potato industry, it now is the world's most important food plant, exceeding both rice and wheat in the number of people it sustains. No other plant in human history has come from relative obscurity to a leading position in world agriculture in so short a time, and the humble potato has all the ingredients of a thrilling success story.

The Spaniards picked up the potato, as they did several other valuable crop plants of this hemisphere, such as corn and tobacco, and carried it to Europe. For the better part of two centuries after its introduction into Europe, it was only a botanical curiosity. Finally someone discovered that the tubers were good to eat. In the cool, moist climate of northern Europe the potato grew wonderfully, and before 1800 it had become a staple agricultural crop. It furnished food for humans, feed for animals, and the raw material for alcoholic liquors, starch, and other valuable products.

The British Isles, especially Ireland, went in heavily for potato culture. Once the plant became established as a staple crop, the Irish lived on it, throve mightily, and increased rapidly. The north European countries grew potatoes in a big way too, but most of them had a diversified agriculture. Ireland depended mainly on potatoes.

Nobody knows when, where, or how *Phytophthora infestans*, the fungus that causes late blight of potatoes, got to Europe. Probably on tubers brought from South America, because it since has been found on wild potatoes there. It began to cause some local epidemics in England before 1840. The spread and growing intensity of the disease are mirrored in the columns of the *Gardeners' Chronicle* between 1835 and 1845, often in Letters to the Editor.

The worse the disease became, the hotter waxed the discussion. Prizes were offered for proof of its cause and for remedies. One contributor advanced the argument that the disease was caused by electricity given off by those new-fangled, smoke-belching, stock-frightening machines called locomotives that ran through the country at a wild speed of twenty miles an hour. Electricity, he argued, was known to have many strange and powerful effects, as Drs. Franklin and Priestley had shown. What more reasonable to assume than that a superfluity of it, generated by the friction of locomotives thundering along at all of twenty miles an hour, should cause plague in human beings and blight in potatoes? He suggested that someone investigate the possibility. The editor cagily made no comment.

When the modest Reverend Berkeley, one of the really out-standing students of fungi in England or anywhere else at that time, reported finding a fungus regularly associated with the disease, and suggested somewhat hesitantly in a letter to the editor of the same journal that said fungus might possibly cause the disease, the editor exposed him to a genteel blast of editorial omniscience.

As it happened, Berkeley was right. To the editor's credit it must be said that before another decade had passed he was publishing a weekly column by Berkeley on the role of fungi in plant disease. It was a somewhat abstruse and wordy column, but then Berkeley had about eleven children whom he supported mainly by his mycological writings, and a certain wordiness must have been justified. He wrote one of the first good books on mycology, in 1858, one from which modern students can still learn a good deal.

In the meantime the fungus had been gradually building up, spreading into new fields, infiltrating into new territory, biding its time, and preparing for Blight Day. A time of rain and muggy weather in July. This came in 1845, and the blight hit with devastating force. Fields that were green and lush one week were dead and brown the next. The tubers were rotted in the ground or, if harvested, rotted quickly in storage. In Ireland the fungus made almost a clean sweep of the potato crop. Famine of the most dreadful sort resulted, with all its attendant horrors. More than half a million Irish died of starvation or of the diseases that accompany starvation. Nearly two million emigrated, chiefly to America. Many of the Irish settlements in the United States can be traced back directly to this 1845 epidemic of late blight.

In 1855 DeBary, a German with a French name, for a time in the pay of the English, went to work on the disease. He soon proved that it was caused by a fungus. This at least removed much of the mystery from it, but enough mystery remained to occupy many researchers for some time. One of the puzzles was how the fungus lived through the winter. Experiments over a long period of time proved to everyone's satisfaction that it did not live in the soil. It did sometimes live over winter as dormant mycelium on stored tubers, but usually such blight-harboring tubers were at least partially decayed, and no sensible gardener or farmer would use partially rotted tubers as seed stock. One of the catches in this was that the fungus, if present on the tubers in the fall, was *likely* to rot them by spring. *Likely* to. Not always. If there was some blight on the potato leaves when the tubers were harvested, spores of the fungus might be washed or scattered onto the tubers. The fungus would rot some of these tubers during winter storage. Some it would infect so lightly that casual or even expert examination would not reveal the infection. If such tubers were planted, the fungus might infect the sprouts, get up above the ground, and there spread to neighboring plants. If the weather was favorable to blight, a local epidemic might be under way.

Until recent times a good many events in the above-described sequence were largely theoretical. Experimental evidence had furnished certain facts, and the spaces between were filled up with logic. To settle the matter, Bonde, a Minnesota Norwegian working in Aroostook County, Maine, planted thousands of tubers known to be infected with the blight fungus. Only about one in twelve hundred of these gave rise to a sprout with blight on it. This single one might be enough to start an occasional epidemic, but it did not explain to Bonde the regular epidemics that occurred in many fields. He thought the fungus must have some other way of overwintering than this relatively haphazard one of remaining alive on tubers, infecting an occasional new sprout, and then spreading out over the fields. He forsook his laboratory for the fields, and in a remarkably short time was able to trace back beginning infections to their source. The explanation of this hundred-year mystery was simple. The blight was coming from potato dumps or cull piles.

Where potatoes are grown in quantity, the seed tubers for next year's planting are stored in warehouses or potato cellars. In the spring, at planting time, the potatoes are sorted over, and usually cut in pieces for planting. The decayed or partially decayed potatoes are thrown out on a pile beside the warehouse. Many of these partially rotted potatoes are infected with *Phytophthora infestans*. On the cull pile the tubers produce a thick growth of succulent sprouts in a hurry. The blight fungus infects these, and by the time the potato sprouts come up in neighboring fields, there are uncounted billions of spores on the cull piles ready to infect them. If the weather is favorable to blight, a local epidemic can start in the neighborhood of every such dump pile, and in a few weeks these local epidemics merge into a regional or national one.

The solution is elementary, as Holmes used to say to Watson. Outlaw such dump piles. Bury, burn, or otherwise destroy the cull tubers. Such laws were passed, in states from Minnesota to Maine. It is one thing to make a law, another to enforce it. Most of those who store potatoes on a large scale also grow potatoes on a large scale, and are alert enough to take advantage of any such public health measure so obviously for their own good. But many a farmer or small-town dweller disposes of rotten potatoes in the spring by merely dumping them out on the nearest available place or putting them into the compost heap. The writer, to his shame, has found potato sprouts infected with late blight on his own compost heap. From such sources the blight fungus can spread by air-borne spores for at least several hundred yards, and in humid, cloudy weather perhaps for miles. The fungus has such a tremendous reproductive potential that once it gets into potato fields and the weather is favorable for rapid increase, it can build up an epidemic in a short time.

Whether one backyard pile of cull potatoes in Maine, New York, Michigan, or Minnesota can infect the whole northern region we do not yet know. It is important to find out, because potato blight is still very much with us. Some of the research involved in finding out looks to a layman like academic boondoggling. Not even a research man would maintain that all research is productive. But the fact that a specific aspect of a research problem, such as catching blight spores from the air, can easily be made to look ridiculous does not mean that it is not vital research, if well done.

The problem is large and complex. This same potato blight fungus can at times also infect tomatoes. In 1946 what started as a small and limited infection of tomatoes in southern Florida became, before the season was over, a general epidemic in commercial tomato fields from Iowa to New Jersey, and had caused a loss estimated at forty million dollars. This may not have amounted to much in our total national economy. But for some of the growers it was more than a minor calamity. On a problem with so many ramifications we need extensive research – in the long run to the interest of consumer and grower alike.

Late blight of potatoes can be controlled fairly effectively by spraying or dusting the plants at the right stage with various fungicides. This involves an outlay of time, money, equipment, and know-how. If blight is not likely to appear, such prevention is an expensive waste. What is needed is an accurate forecasting service so that growers in a given region can be informed by radio when blight is imminent, and so can apply preventive fungicides only when necessary. This forecasting is now being tried out in several different countries. After more research, it may become a regular and integral part of all large-scale potato culture, and another one of our major public enemies among the fungi will be a has-been.

Considerable progress has been made in the breeding of blight-resistant potatoes too, but this is not a short and simple process. And, as would be expected, the blight fungus is not without some tricks of its own. It is a wily opponent. Research will enable us to keep ahead of it, not eliminate it. Sanitation in the matter of eliminating potato dumps, the development of resistant varieties, the use of fungicides when and where necessary – all help to keep late blight in check. As a result blight takes an annual toll of only 2 or 3 per cent of the crop. That is still many times more, in cash, than it costs for research to eliminate this elusive enemy.

Additional Reading

E. J. Butler and S. G. Jones, *Plant Pathology*. London: Macmillan & Co., 1949. 979 pp.

James G. Dickson, *Diseases of Field Crops*. New York: McGraw-Hill Book Company, 1947. 429 pp.

Bernard O. Dodge and Harold W. Rickett, *Diseases and Pests of Ornamental Plants*. Lancaster, Pa.: Jaques Cattell Press, 1943. 638 pp.

Mayo Foundation, *Lectures on Plant Pathology and Physiology in Relation to Man*. Philadelphia: W. B. Saunders Company, 1928. 207 pp.

John Charles Walker, *Plant Pathology*. New York: McGraw-Hill Book Company, 1950. 699 pp.

CHAPTER 6

Fungus Parasites of Plants: Rusts, Smuts, and Heart Rot

FREEDOM from famine would hardly have been possible without research on the fungus diseases of our economic plants. If it has resulted in the overproduction of potatoes or wheat in the last decade, and the consequent economic ills, these troubles are minor ones compared to those resulting from so recent an epidemic of wheat rust as that of 1916. At that time the writer was a young and comparatively inoffensive boy in North Dakota, one of the breadbaskets of the world, a major producer of wheat. In the winter of 1916–1917 the writer lived mainly on corn-meal mush. The famous breadbasket was empty, that year. As it had been in 1904 and was to be again later. Because of stem rust. Stem rust deserves a book of its own, and the discussion of it and related rusts that follows must leave out even many essentials, let alone details.

Rust Fungi

The rust fungi are not very numerous as fungi go. There are only from about three to four thousand different species of them. But in economic significance and in biologic complexity they overshadow some of the more numerous kinds. A brief summary here of their lives and habits will give the reader a general basis for understanding how they operate.

First, all rusts are obligate parasites; they will grow only in the living tissue of living plants. Second, most of them produce several different kinds of spores, in regular and almost unalterable succession. There are thousands of different fungi that regularly produce two kinds of spores, one mostly during the growing season that facilitates rapid spread and increase, and a second during the fall that serves to overwinter and start the fungus off again in the spring. The technical name for this production of different kinds of spores is *pleomorphism*, which means *many forms*, or *different forms*.

Among the fungi the rusts have carried this pleomorphism to extremes. Many of them produce five different kinds of spores during their annual life cycle, each in its own unalterable place and sequence. One kind of spore may serve only as gametes — sex cells comparable to egg and sperm cells among the animals. These are the pycniospores. Usually they appear in early spring. After sexual fusion has occurred, aeciospores are produced, each aeciospore containing two nuclei, one from each of the pycniospores that fused shortly before. The aeciospores are produced in some quantity, and are carried by the wind to the appropriate host, which they infect. They germinate, establish a mycelium in this host, which mycelium within a short time gives rise to a crop of urediospores. These are mostly "repeating" spores. They are carried to other plants of the same host species, germinate, infect the host, produce a mycelium within the host, and this produces more urediospores. Successive crops of urediospores may be thus produced during the growing season, each crop serving to spread and intensify the rust. After a number of crops of these repeating urediospores have been produced, the same mycelium that gave rise to them will begin to produce teliospores. These either germinate at once or lie dormant over winter and then germinate to produce sporidia or basidiospores. The two nuclei that originally came from the two fusing pycniospores have remained together, dividing together whenever a new cell of mycelium was formed or whenever a new aeciospore, urediospore, or teliospore was formed. In the mature teliospore these fuse, then quickly undergo reduction division, and form four haploid or 1x or 1n nuclei, typical of many germ cells throughout the plant and animal kingdoms. One haploid nucleus enters each basidiospore. The basidiospores are carried by the wind to the appropriate host, germinate, infect it, and develop a sparse mycelium, which produces pycnia. Within these pycnia pycniospores are produced, each of them with a single haploid nucleus. Pycniospores of opposite sex fuse to give rise to aecia and aeciospores.

It sounds a bit complex, and it is. The diagram that follows (Figure 6) may help clarify it. The writer recognizes that the layman or beginning student confronted with this rather involved life cycle of a typical rust fungus cannot be expected to grasp all the details of it at once, without the aid of laboratory specimens, nor is it intended that he should. But if nothing else, at least he can recognize that the rust fungi are in some ways rather complex organisms whose mode of life differs from that of the ordinary plants with which he is acquainted.

A further characteristic of many rust fungi is that during their life cycle they parasitize two entirely different, distinct, and unrelated host plants or groups of host plants. Normally the pycnia and aecia are formed on one host plant, and the uredia and telia upon another. As examples of this, the common black stem rust of cereals forms pycnia and aecia upon the leaves of barberry (without doing the barberry any appreciable damage) and uredia and telia upon the stems and leaves of cereal plants (doing the cereal plants sometimes great harm indeed). White-pine blister rust forms pycnia and aecia within the bark of white pine, and the aecia disrupt the bark to such an extent that eventually the tree is killed; it forms uredia and telia upon the under sides of the leaves of currant and gooseberry bushes, and seldom does these much harm. On the other hand, flax rust infects only flax; it does not have an alternate host. The same thing is true of hollyhock rust; it infects only hollyhock, and produces only pycniospores and teliospores — the bare minimum of spore types to get along.

Obviously the rust fungi are not static, in an evolutionary

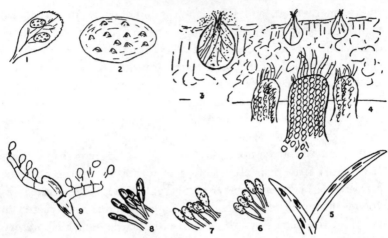

Figure 6. Life history of the stem-rust fungus. (1) Infection spots on a barberry leaf, each of which has developed from mycelium derived from a single basidiospore. (2) A single infection spot on a barberry leaf, with the tips of numerous pycnia projecting above the surface of the leaf. (3) A section through a single pycnium. Bristle-like branches grow out from the inner wall of the pycnium and protrude from the opening. Masses of pycniospores are produced within the pycnium and exude from the opening in a sticky fluid. (4) Pycnia have been fertilized, and aecia have developed on the under side of the leaf. Aeciospores are shot out, to be disseminated by the wind. (5) Pustules of rust on the leaves of a wheat plant, resulting from infection by aeciospores. (6) Urediospores, produced in the pustules just shown. (7) The urediospores "repeat," and produce successive crops of urediospores during the growing season. (8) Teliospores, the "black rust" stage, developed in the fall. These lie dormant over winter. (9) A teliospore germinating. Each of the cells gives rise to a promycelium, or basidium, on which four basidiospores are produced. These are forcibly discharged, and infect the leaves of barberry, giving rise to the infection spots shown in *1.*

sense, but dynamic. They are exploring all possibilities within their range. One would think that a fungus which operated as an obligate parasite, which produced several kinds of spores in unalterable succession, and which infected the leaves of fir trees in one part of its cycle, and the leaves of certain kinds of ferns in another part of its cycle, or which infected pines and gooseberry leaves, or which infected wheat and barberry, would be severely penalized from a survival standpoint. The cycle seems unnecessarily complex, and akin to that of certain protozoa and certain insects which also pass different portions of their life cycle in different and unrelated hosts. The best evidence to indicate that they have not penalized themselves unduly by this complex arrangement is that they not only survive, but thrive. Before modern man arrived on the scene, many of these rust fungi must have been relatively inconspicuous members of the fungus flora, even as many of them still are upon our wild plants. But when we began to cultivate certain plants on which the rusts spent the major portion of their life cycle, the rusts became important to us. And they probably will continue to be significant factors in our plant economy for some time to come. A better idea of what they mean to us, what they do, and how they do it, will be got from a description of several typical rusts.

COFFEE RUST

Coffee, almost as essential to our modern civilization as tobacco and television, is a relative newcomer as an economic plant. If it was grown in ancient times, it was not grown extensively. It probably is native to Abyssinia, and it has been cultivated for only a few hundred years. As a beverage, coffee became popular in England during the time of Samuel Johnson, when the coffeehouses first arose, and those who frequented them not only drank coffee but traded in coffee and other commodities. Eventually it was found that the coffee tree could be cultivated successfully in Ceylon. Plantations were established, they increased in number and size, and by 1850 coffee was the economic mainstay of Ceylon.

Sometime between 1850 and 1870 coffee rust appeared. It probably had been present before, but in such small amounts as to go unnoticed. It was called to the growers' attention only when it began to kill not just individual trees but entire plantations. At that time only a few people knew that fungi were living things, and still fewer knew that they could cause plant diseases. Between 1870 and 1880, by which time plant pathology as a study had developed at least to the swaddling-clothes stage, a plant pathologist was called out from England to Ceylon to study this mysterious coffee blight. He found that it was caused by a fungus, not by any of the many other things that had been suggested as its cause. He was unable to find any workable cure for the disease. Within the next twenty years coffee as an economic crop was eliminated from Ceylon, and many of the large plantations, and their stockholders, were ruined. Coffee cultivation moved to the Western Hemisphere — without the rust — and the crop has become an essential part of the economy of some of the South and Central American countries. Other diseases plague it there, but at least these do not totally restrict the cultivation of the coffee tree, as the rust did in Ceylon.

WHITE-PINE BLISTER RUST

White-pine blister rust is caused by a fungus that apparently is native to those portions of Siberia that support the growth of *Pinus cembra*, one of a number of "white" pines. As a group, the white pines are recognized by having their needles in clusters of five. White pines are of little importance in Europe or Asia, but several species are among the most important and valuable timber trees of the United States. The eastern white pine once made up a major part of the upland conifer forest from Maine to Minnesota; the western white pine, still more valuable, grew from western Montana to the Pacific, and the sugar pine covered the slopes of California and Oregon. Most of the easily accessible original stands of these were cut as the country was developed, to be sure, but they were cut for economic use. They helped build the country. Many of these original stands have been replaced by natural means or by reforestation, but white-pine blister rust has made this replacement a difficult and expensive and sometimes hazardous business.

White-pine blister rust was introduced into the eastern United States on seedling white-pine trees that had been raised in France and sent to the United States. In the early days of forestry in the United States, it was assumed that we did not have the necessary know-how to raise seedlings of our most valuable forest trees. So we shipped seeds to France or Germany; the seedlings were raised there, then sent back to us for planting, along with whatever fungus and insect pests had become established on them. White-pine blister rust was known in Europe long before 1900 as a destructive parasite of American white pine that had been planted there, but this did not prevent the importation into the United States of seedlings infected with the disease. This rust was first found in the United States in white-pine plantations in New York, shortly before 1910. It was later found in another plantation, on Vancouver Island, off the west coast, adjacent to some of our best stands of western white pine. From these two sources it spread over all the range of both the eastern and the two western species of commercial white pines in the United States. In regions favorable to the development of this rust, it has caused heavy losses. But these losses, heavy as they have been up to the present time, are relatively unimportant as compared with the future.

This rust fungus, like many another rust fungus, has an alternate host. In this case the alternate host is Ribes — wild or cultivated currants and gooseberries. In most of the regions where white pines are an important or economically significant tree, Ribes bushes are present in abundance. The only practical method yet found of controlling this particular rust fungus is to eliminate the wild currant and gooseberry bushes within and around the stands of white pine. In some parts of the country this is a relatively inexpensive procedure. In other parts, especially in the western United States, it is difficult and expensive, justified only where white pine supports the economy of the region and where the growing of white pine would be impossible without it. Most of the expense of this control of white-pine blister rust has been borne by the federal government, but since private timber companies also contribute to it, sometimes fairly substantially, it may be assumed that by big business also control is considered essential in the long-time management of those forests where white pines make up an important part of the stand.

STEM RUST OF CEREAL PLANTS

Stem rust, or black stem rust, so called to distinguish it from several other rusts that infect our major grain crops, has been an important disease of several kinds of cereal plants ever since these were first cultivated. All of the various civilizations that have developed in the world have been based to a large extent on one or another of the cereals. Rice in the Orient, corn in Middle and South America, barley and wheat around the Mediterranean. The culture from which we developed has been mainly a wheat culture up to very recent times.

Evidence in the Old Testament indicates that failure of the wheat crop was no uncommon cause of famine then. We cannot prove that rust was a major factor in these crop failures, but since both wheat and stem rust apparently arose in this region, it is only reasonable to suppose that rust was at least sometimes involved. That rust caused heavy losses in olden times is attested to by the fact that the ancient Romans established a god of rust and sacrificed to him to keep this plague from their fields. Wheat was a staple of the Roman economy. In the United States Department of Agriculture yearbook *Climate and Man* the climatologists tell us that the first three centuries A.D. were times of unusually high rainfall along both the north and south shores of the Mediterranean, where civilization was most concentrated. In other words, rust years. Two or three rust years in succession in what was then the breadbasket of Western civilization could have led to widespread starvation, to epidemics of cholera, typhus, and other diseases that accompany starvation, to a breakdown of established rules and customs, and a general outcry against the government. We have, of course, no direct means of measuring the effect of plant rust upon the economy of those times, but it may have played a part in the recurrent food shortages of which we know, and the inference is perfectly reasonable that it was one of the multitude of factors in the social unrest and turmoil accompanying the decline of the Roman Empire. The fact that historians have neglected this and many other biological aspects of history does not mean that biology has been unimportant. Few historians, ancient or modern, have had even a speaking acquaintance with microbes, much less an appreciation of the influences of these microbes on the fate of the civilizations they study.

As wheat became a major crop in the Great Plains region of the United States and western Canada and was grown over an area of thousands of square miles, stem rust became a potent factor in determining the amount and quality of the wheat harvested. Quoting from Butler and Jones: "It is recorded that in the three North Central States of Minnesota, North, and South Dakota, and in Western Canada, the losses in these two areas in one year amounted to 180 and 110 million bushels respectively; and in Manitoba and Saskatchewan, during 1925 to 1935, the average annual loss was over 35½ million bushels (11 per cent of the possible yield), a monetary loss estimated at 30,784,000 dollars." * The loss in bushels or dollars tells only a part of the story: for many people in the regions concerned it meant black despair, financial ruin, and a bitter struggle to survive. There is almost no major wheat-growing region of the world that has not, in the last fifty years, experienced a series of heavy crop losses from this rust fungus.

A number of characteristics and habits make this rust both difficult to live with and difficult to get rid of. The major ones will be summarized briefly.

Hosts and life history. Puccinia graminis, the fungus which causes black stem rust, has two hosts, or groups of hosts, on each of which it spends a part of its life cycle. In the spring it produces pycnia embedded in the upper sides of the leaves of barberry (and a few close relatives of barberry). Soon after the pycnia are ripe and their spores have mutually fertilized one another, aecia are borne on the under sides of the leaves. Within

* E. J. Butler and S. G. Jones, *Plant Pathology* (London: Macmillan & Co., 1949).

these aecia aeciospores are produced in large numbers. They are dry and powdery, and when they are ejected from the aecia, they are carried by the wind for hundreds of yards, or, as we have already seen in Chapter 2, in some cases, hundreds of miles. These aeciospores can infect wheat, oats, barley, rye, timothy, redtop, and some wild grasses, all of which together make up the other group of hosts of the fungus. An aeciospore that lands on the leaf or stem of a young wheat plant germinates, infects the tissues of the plant, and produces a patch of mycelium within this infected tissue. Within about two weeks this mycelium produces a crop of rust-colored spores, the urediospores. These are able to "repeat" — they blow to other wheat plants, germinate, infect the host plant, and within about two weeks produce another crop of urediospores.

This can continue all through the growing season, and, what is worse, in continental North America it can go on from year to year without much interruption. The urediospores that live through the winter on winter wheat in southern Texas or northern Mexico develop and increase in that region early in the year. As the wheat ripens northward, the rust moves northward with it. In the fall, when the wheat in the northern United States and the adjacent region of Canada is harvested, the urediospores may blow south, to infect the winter wheat just coming up there. This is not mere speculation. For somewhat more than thirty years the plant pathologists in the region concerned have been tracing the spread of these air-borne urediospores. They know that sometimes urediospores picked up by the wind in Kansas, Nebraska, or Oklahoma may be carried northward to South Dakota, North Dakota, Minnesota, and Manitoba, to shower down in hundreds or thousands per square foot. If this occurs at the right time, and if the subsequent weather is favorable, the result may be a widespread outbreak of rust.

It may be. It doesn't occur inevitably or always. We often have such spore showers without any subsequent appreciable damage from rust. A number of things (mainly temperature, humidity, and the stage of development of the plant) have to combine just right in order for the rust to get established and to increase rapidly just after the spores have arrived. Sometimes this happens over a large area, and we have a general epidemic. Sometimes there is such a general spore shower, but weather conditions favorable to the development of rust subsequently prevail only over a few local areas, and in these the rust develops. These urediospores can survive the winter in the north, but those which do so, germinate in early spring before any grain is present to infect and so cause no harm.

In the late summer or fall the same mycelium that has been producing urediospores begins to produce the black teliospores. These lie dormant over the winter. In fact, they require weeks of cold weather, followed by several periods of alternate wetting and drying, before they will germinate. They germinate over a period of several weeks in the spring, just when the barberries are putting out leaves, and they produce a fifth type of spores, the basidiospores or sporidia, which infect the leaves of barberry.

The main damage done to the grain is a mechanical rupturing of the outer skin, or epidermis, of the infected leaves and stems by the urediospores. If this occurs early enough and in sufficient amount, the vital juices of the plant are lost, and the seeds shrivel.

Varieties and races of the rust fungus. When the life history of *Puccinia graminis* was first worked out, shortly after 1850 (by Anton DeBary, the same man who discovered the cause of late blight of potatoes, a strictly grade-A genius to whom our present world is obligated in more ways than we realize and who with justice is known as the father of modern plant pathology), it was thought that *Puccinia graminis* was just *Puccinia graminis* — that it was a single uniform species able to infect any one of its several grass hosts with equal facility. That one could, for example, take spores of the rust from rye and infect wheat with them, or the spores from wheat and infect rye with them. It is not so simple as that. By the 1880's some rather painstaking and extensive cross inoculation — taking the repeating spores from wheat and inoculating rye, barley, wheat, oats, and other grasses with them, and doing the same with spores from each of the several different hosts — had shown that the rust comprised several different and distinct varieties. They all looked alike, but differed in their ability to infect different host plants. The variety from wheat, named *Puccinia graminis tritici*, infected wheat mainly, barley and rye and some wild grasses only slightly. The variety from oats, *Puccinia graminis avenae*, infected oats mainly, some wild grasses slightly, and wheat, rye, or barley not at all. All of these different varieties produced their pycnia and aecia on the barberry, sometimes on the same leaf.

This was only the beginning. Still later it was found that the variety *Puccinia graminis tritici*, whose principal host was wheat, was by no means a uniform variety, but comprised many different races, one race of it able to attack certain varieties of wheat, another able to attack other varieties. Over two hundred such parasitic races have been described, and over the last thirty-five or forty years, the distribution and prevalence of the principal ones in some of the principal wheat-growing regions of the world have been charted. Some of these races are relatively innocuous, some are actual or potential killers. Obviously the existence of a large number of different parasitic races, each differing from the others in the varieties it can attack, the conditions under which it will thrive, and the damage it can do, complicates the breeding of wheat varieties resistant to black stem rust.

Getting down still finer, within the last ten years it has been found that some of these individual races of the variety *Puccinia graminis tritici* are not uniform and homogeneous. As an actual example, it is now known that the race known by the number 15 (it was the fifteenth race of *P. graminis tritici* identified) consists of the standard 15 and of a variant of it known as 15B. Most of the wheats we now grow are resistant to race 15, and so we do not have to worry about the standard race 15. But many of the wheats we grow, including durum wheats and winter and spring bread wheats, are very susceptible to race 15B. Race 15B, discovered only in the last twenty years, and for a long time found only in the vicinity of barberry bushes, now is on the loose. In 1950 it caused rather heavy losses, particularly on durum wheats, in the northern portion of the spring wheat region of the United States, and we can anticipate more trouble from it in the future.

Where do all these varieties and races come from? As early as a century and a half ago the intervention of the barberry bush was at least suspected. Shortly after 1800 a Danish school teacher proved that rust from barberry leaves could be transmitted to rye plants. Nobody paid much attention to his work, partly because it was not widely publicized, partly because he was only a high school teacher and not a "scientist." He had not paid his dues to the scientific fraternity; he did not have the handshake — a form of snobbery that is by no means a thing of the past. But this man did uncover the secret of the stem-rust fungus, even though much more work was required to bring it into the open. About 1930 it was found that the pycniospores of rust fungi, previously thought to be functionless, actually functioned as sex cells, or gametes. Different races and different varieties of the fungus can and do hybridize on the barberry leaves. The barberry is essentially a plant-breeding station for the rust fungus. It is almost as if the rust fungus had anticipated man's attempt to eliminate it by the production of varieties of grain resistant or immune to rust infection, and had in self-protection developed its own program of breeding new and different races able to attack these new varieties.

Control of black stem rust. Once the barberry was known to be the alternate host of black stem rust, programs were got under way in various countries aimed at the elimination of the barberry. In much of Europe where wheat is an important crop, the barberries have been eliminated or their numbers reduced to such an extent that stem rust no longer is a major problem there. In the United States, over the past few decades, the barberries have been eradicated from most of those areas where spring wheat is grown extensively. This has eliminated the regular local epidemics that previously were so common. It could not eliminate those epidemics that arose from the repeating urediospores working northward over the Great Plains area. Rust is too wily and resourceful an enemy to be beaten by any single attack, but the eradication of barberries was and still is an essential campaign in our war against it.

Another essential campaign is the breeding of varieties resistant to the fungus. This work got under way in Canada about 1900, and in the United States shortly after that. One of the first products of this program was the variety Marquis, which was released in 1913 and soon came to occupy a tremendous acreage in the spring wheat area. At first it was exceedingly resistant to rust, or rather to those races of rust which at that time were most prevalent. As soon as Marquis became generally grown, other races of rust that had previously been present but inconspicuous, but which were able to attack Marquis, began to increase; and before long these practically eliminated this valuable variety. Marquis was largely replaced, in the twenties and thirties, by several other rust-resistant varieties. These in turn eventually were attacked by still other races of rust that previously had been present in only infinitesimal amounts, and the rust-resistant varieties of the thirties have been replaced by still newer varieties. Thus the breeding of varieties resistant to rust is not something that can be carried on for a time and then dropped, as once was believed. *Puccinia graminis* will be around for some time to come. The widespread and regular epidemics of the past are getting fewer and farther between. In the Great Plains area of the United States, rust-resistant winter wheats in the South, rust-resistant spring wheats in the North (and also in the northern half of Mexico and the southern half of Canada), the elimination of barberries on which the fungus can breed races to attack these new varieties — all have been and still are important in putting stem rust on the run. Work by one or a few research men in one or a few research laboratories does not suffice here. The attack has to be made on a national and international basis, and that is now being undertaken.

Smut Fungi

The smut fungi were so named because they produce masses of dark, smutty-looking spores on the plants they infect. In nature most of them are almost obligate parasites, but they can be grown on artificial media in the laboratory. There are only about nine hundred species of them, but a number are common and destructive parasites of various economic plants; so they have received considerable study. To illustrate the fecundity of fungi in producing spores, *Ustilago zeae*, which causes smut of corn, and *Tilletia tritici*, which causes bunt, or stinking smut of wheat, were mentioned in Chapter 2. We shall now look at *Tilletia tritici* once again, and take up also another common variety of the fungi that result in smut disease.

COVERED SMUT OF WHEAT

The fungus, *Tilletia tritici*, that causes covered smut of wheat forms its spores within the pericarp, or covering skin, of wheat seeds and fills the seeds with a powdery black mass of ill-smelling spores. These spores are released when the wheat is harvested and threshed; they may land on healthy seed or be carried by the wind for many miles. If seed contaminated with these spores is planted, or if noncontaminated seed is planted in soil that is contaminated with them, and if conditions are favorable for the fungus, the spores germinate when the seeds do. Germ tubes of the fungus infect the growing point of the stem, and mycelium, having grown up with the stem, fills the seeds with spores when the seeds are formed. Some of the plants infected with the fungus will be stunted. Not only does the fungus reduce the yield; if the harvested seed is seriously contaminated by it, an unpleasant odor and flavor are imparted that reduce the quality of the grain for flour. Before the days of modern grain inspection and modern grain cleaning, smutty bread must have been too common for comfort. Covered, or stinking, smut of wheat was recognized as a serious pest in Europe several hundred years ago, and it is likely that the undesirable flavor of bread made from such smutty grain was more obvious to people than any actual reduction in yield, which sometimes is a subtle thing.

Tillet, a Frenchman, in 1753 published the results of extensive experiments which to him proved that this disease was caused by a "contagion" carried on the wheat seeds. Treat the seed with something to kill the contagion, and smut would be reduced or eliminated. Treat it with what? Tillet had a classical education, with some respect for ancient tradition, and so he tried out the old remedies of steeping the seed in sheep manure, cow manure, pig manure, spirits of niter, lime, and a few other things, but mostly manure. He decided that the best preventive in which the seeds could be steeped was putrified urine. Shades of Aristotle! He did some remarkable work, but certainly his recommendation of putrified urine as a seed treatment was no milestone in the new science of plant pathology. In the matter of the causal nature of bunt or stinking smut, he was getting warm, but he did not clinch it. The most important link in his chain of evidence was missing. This was supplied by another Frenchman.

Benedict Prevost began life in Geneva, Switzerland, about the time of Tillet's publication on bunt. He became successively an apprentice in engraving, an apprentice in commerce, a private tutor, a mathematician with leanings toward natural science, and finally a professor of philosophy.

He worked on bunt, and identified the bunt itself as fungus spores. He sowed these spores in water and watched them germinate, and was the first man consciously to observe and clearly to describe the germination of fungus spores of any kind. By many experiments over a period of ten years he proved that this fungus was the direct, immediate, and sole cause of covered smut, or bunt, of wheat, and he so stated, clearly and concisely, in the year 1807.

In the course of his studies he observed the effect of time, temperature, age, and many different chemicals on the germination of these bunt spores, and the relation of soil, weather, and other environmental factors to infection by the disease and to its development. His whole approach to the problem as well as his methods of investigation was modern. He knew that the mystery of stinking smut could be unraveled only by accurate research on the fungus itself, not by a study of ancient and erroneous speculations on the nature of disease. His mental habits as well as his

Plate 5. Mycelium of a fungus and many colonies of yeasts growing from the interior of a flour beetle.

Plate 6. Fruit body of a wood-rotting fungus, *Polyporus dryophilus*, on the trunk of a white oak tree. After the fungus has grown in the tree for 25 to 50 years it produces its first fruit body such as this, and subsequently may produce another every year or so for decades or centuries. A quarter of a century after the photograph was taken, the tree still appeared vigorous; it doubtless will survive for another 50 or 100 years.

Plate 7. An attractive clump of fruit bodies of *Pleurotus ulmarius* growing from a knothole of a decaying trunk of box elder. This fungus is a common cause of decay in trees planted along boulevards.

actions are well illustrated by a brief quotation from his work on this disease:

"Having proved, as we have seen in the preceding section, that an infinite number of substances when placed in water give it the property of destroying the germ of bunt spores, it remained for me to investigate which of these substances combine in the highest degree all of the qualities of a preventive applicable to agriculture. I had begun to proceed methodically towards this, when a fortunate incident considerably shortened this work.

"Having washed some bunt with several changes of well water, then water that had been distilled in a large copper alembic, and having left it for some time in a glass goblet containing this water, I placed drops of the water from the goblet, containing several hundred gemmae or spores of bunt, in a watch glass half full of highly purified distilled water. To my great astonishment, these gemmae either did not germinate or germinated very poorly, whereas some others, in similar circumstances save for the copper, germinated as usual. I decided then to direct my researches first and principally toward copper and the copper salts."

So well were his researches directed that within a few growing seasons he had devised effective and practical treatments, consisting of dipping the seeds, just before planting them, in a solution of copper sulphate or copper carbonate. The value of these treatments soon became known. They were widely used in France and England within a short time. With some improvements and changes, they were used till after 1900, and they formed the basis of some very effective seed treatments for other diseases also. It is true that some growers even before Prevost's time steeped their seeds in milk of lime brewed in a big copper kettle, and in this way reduced the amount of bunt in wheat, or so they said. But Prevost's discovery was made independently, and in any case he was the first to prove that copper salts were good fungicides (for some fungi) and to test and recommend specific doses.

Although farmers adopted his recommendations, scientists ignored his work. Prevost had no formal training in any botanical field, and little enough formal schooling of any kind. He had no standing among the academicians of his time and region. Had he been a good self-advertiser, he might have gained recognition; being of a quiet and retiring nature, and a lover of truth rather than argument, he chose to let his work speak for itself. But he was too far in advance of his time. His results were contrary to what was taught wherever anything concerning fungi or plant disease was taught, which was not in many universities. Most of his experiments were simple, direct, and to the point; his language was equally so. His technics were not involved. Anyone with a reasonable mastery of the microscope and of simple chemistry could have repeated his work. But professors, after long years of teaching, are apt to confuse their lecture notes with the fountain of truth. It is an occupational disease. The professors of the day who could have won permanent renown by merely espousing Prevost's work preferred the security of tradition and gave him the silent treatment. When Prevost saw how the wind was blowing, he gave up his researches with fungi and plant disease, and accepted a chair of philosophy in a Protestant school.

Two other names should be recorded in the fight against *Tilletia tritici*. In the latter 1800's Oskar Brefeld, in Germany, followed the fungus through the wheat plant, grew it in artificial culture on agar, and uncovered several secrets important to our understanding of it, among them that sometimes infected plants were so greatly stunted that they might not even produce seed. Early in the present century McAlpine, in Australia, devised a copper carbonate dust much superior to the older seed treatment. Then the organic chemists discovered the field of fungicides, and in the last thirty years several very effective seed treatments have been devised. In fact, for a time it looked as if a combination of seed treatment and the breeding of varieties resistant to the disease had eliminated covered smut as a serious problem.

But this fungus also turned out to have some tricks that were not apparent at first. A few years ago dwarf bunt turned up in the western United States, where in some areas it reduced the yield almost to the vanishing point. It proved to be a new variety of *Tilletia tritici*, and one not amenable to control by seed treatment. The spores were able to survive in the soil for years, and the usual seed treatments had little or no effect on them. This fungus, as well as other smut fungi, shares the biological benefits inherent in sex, and it exploits these benefits as vigorously and as ingeniously as do the rust fungi. The species *Tilletia tritici* is made up of many different parasitic races, and new races probably are appearing all the time, even as new races of it have been bred experimentally in the laboratory. Though we have achieved fairly effective control of this disease throughout most of the main wheat-growing areas most of the time and are keeping ahead of it, it still is definitely in the race, at least in certain regions.

LOOSE SMUT OF WHEAT

Ustilago tritici, the fungus that causes loose smut of wheat, has a mode of life somewhat different from that of its relative just described. Here the spores infect the flowers of the wheat plant and germinate; mycelium invades the germ of the developing seed and lies dormant within the seed during the winter. One cannot distinguish, by ordinary examination, sound seed from seed infected with this fungus, although the presence of the fungus can be detected by microscopic examination, if one knows where and how to look for it.

When such infected seed is planted and germinates, the fungus begins to grow. Its mycelium grows up with the growing point. When flowers are formed, these are converted into a mass of spores of loose smut. These are carried about by the wind, they infect the flowers of neighboring healthy plants, and in this seed the fungus lies dormant until the seed germinates.

Since the mycelium of this fungus is present *within* the dormant seed, it cannot be reached by fungicides applied to the outside of the seed. It can be controlled by steeping the infected seed in hot water, but the temperature necessary to kill the fungus is only a few degrees below that which will kill the seed. Unless expertly done, this heat treatment either does not kill the fungus, or it kills both the fungus and the seed. So in modern times the most effective control has been the breeding of varieties resistant to attack by the fungus.

As probably would be expected by those who have read the above, this fungus also comprises a large number of different parasitic races, and new ones are continually being produced by hybridization on the part of the fungus. Also, the problem is not simply one of breeding a variety of wheat resistant to loose smut. That would be easy. The variety must be resistant to covered smut as well, and to stem rust, leaf rust, scab, root-rot, and mildew. In some regions it must also be resistant to Hessian fly and to other insects. It must also be adapted to the soil and climate of the region in which it is grown, and possess various other essential agronomic characters. And, of course, if it is a wheat that is to be milled into flour and baked into bread, it must have the milling and baking qualities required by those industries; otherwise it will not be marketable. The same applies to those wheats that are to be used for cake flour or to be processed into macaroni and spaghetti. Since the population of each of the fungi that cause the diseases listed above is made up of a large number of different parasitic races, races that differ from one another in parasitic capabilities and potentialities, and all of which are constantly producing new races by hybridization, mutation, adaptation, or combinations of these processes, the breeding of varieties resistant to them all is a big job. We are getting ahead of some of these fungus parasites, but it will be a long time before we shall be far enough ahead to be complacent about our advance.

Fungi That Cause Heart Rot in Trees

Among the many fungus diseases to which trees are subject, the decays, or rots, of the trunk offer an interesting example of a group that differ in some ways from those discussed above. All trees are subject to decay. The normal course of events in nature is for trees to grow up and form a stem or trunk of greater or lesser size which increases in height and diameter from year to year. The outer portion of this trunk consists of wood in which many of the cells are alive. This is known as sapwood. The inner portion of the trunk, usually dark in color and with few or no living cells, is known as heartwood. This inner and dead heartwood is the favorite place for wood-rotting fungi to grow. They rot the wood and consume it, the decayed tree topples over, and in years or decades its shell is converted into humus and soil to nourish succeeding generations of trees. Some kinds of trees, like the forest aspen and balsam fir, are short-lived primarily because they are very susceptible to invasion by wood-rotting fungi and, once invaded, are rapidly decayed. Redwoods are long-lived partly because their heartwood is very resistant to decay. Resistant, not immune. Many of the redwoods almost a thousand years old have had wood-rotting fungi growing in them for a goodly portion of that thousand years.

The fungi that decay the interior of our trees enter them in various ways. Through unwounded roots, through wounded roots, through stubs exposed when branches die and fall off from decay, through frost cracks, fire scars, blazes made by woodsmen and boy scouts, through the holes drilled in the trees by foresters with increment borers (drilled to find out how old the trees are and how fast they are growing), and through a multitude of other natural and artificial wounds.

Once inside the heartwood, these wood-rotting fungi grow relatively slowly — from a few inches to a few feet a year. As they grow, they digest the wood. Some begin at the roots and grow upward. Some begin at exposed branch stubs and work both up and down. All of them eventually consume most of the heartwood. Some of them continue on out to the sapwood, go through it to the bark, and kill the bark. A standard question in Ph.D. examinations used to be: "Is a wood-rotting fungus growing in a tree a saprophyte or a parasite?" Whatever the candidate answered, he was wrong. There are some good parasites among the wood-rotting fungi, and some do kill trees. But most of them merely decay the dead wood. Except for weakening the tree mechanically, and consuming the wood, they do no harm. Thus a wood-rotting fungus in a shade tree does not necessarily spell the early doom of that tree. Depending on the fungus and the tree, such an infected tree may still outlive its owner and his sons. See Plates 6 and 7.

In forest trees the major protection against loss from such rot must necessarily be an indirect one. If trees are subject to rot at an early age, they are utilized before rot becomes of major importance, or they are gradually replaced with other species more resistant to decay, or some use is found for the partially decayed wood. In shade trees, what has come to be known as tree "surgery" has been resorted to. In this, the aim is to remove the decayed wood and then fill the resulting cavity with some substance firm enough to allow the tree to grow over it and thus cover the wound made by the "surgeon" to get into the tree. Actually, there is no experimental evidence to indicate that such surgery will add to the length of life or vigor of the tree.

Most wood-rotting fungi in trees extend for from several inches to several feet beyond any visible decay. Removing the decayed wood merely removes that wood with which the fungus is mostly finished; it does not remove the infection. Filling the resulting cavity with the compounds mainly used for this purpose serves to keep the wood moist and to maintain conditions favorable for further penetration by the fungus left in the wood. It gives the owner some satisfaction to have done something for his tree, and the typical suburbanite seems to feel almost as though he had an estate once he has had a tree or two filled. Other than this, surgery probably is of little use. Some years ago it was stated in the Proceedings of the American Shade Tree Conference, a publication by those whose business is tree care, that the quality of tree care in any region was inversely proportional to the amount of cavity work done. There has been a considerable amount of hocus-pocus in this aspect of tree care, and there is some still. In its present stage of development the excavating and filling of trees to prevent decay is hardly worth serious discussion. But the writer recognizes that this will not deter those who have trees they value from having them excavated and filled, in the belief that thereby they are adding years to the life of the trees and possibly storing up some credit for themselves.

As an indication of how little work has been done in this general field, the matter of fungicidal wound dressings for trees is worth mentioning. When branches are removed from trees, it has been customary to paint the resulting exposed wound with what is called a wound dressing. In the olden days a mixture of wood ashes, cow dung, and putrified urine was used, the idea being that the worse the dressing smelled the better it would be. In the last fifty years various paints have been used, including the common oil paints used to cover the outside of buildings. No one thought to test such paints as fungicides until a decade or so ago, and then it was not in connection with this sort of work. Ordinary paints are not fungicidal; in fact, far from their offering the fungi any particular discouragement, many fungi grow in them vigorously and with relish.

Then asphalt paints were used. They had a sufficiently disagreeable odor to make them seem fungicidal. Shortly after 1940 a fungus disease cropped up on plane trees, or sycamores, in the eastern United States, and caused a good deal of damage here and there among the trees commonly planted along streets. It soon became apparent that this disease was prevalent almost only where tree surgeons or others were using asphalt wound dressing to cover wounds exposed when branches were removed. Samples of asphalt wound dressing from their buckets were streaked on agar plates, and the fungus was found to grow vigorously from it. The "surgeons" were not protecting the trees, they were inoculating them with the fungus they were supposed to be protecting the trees against. A research worker in the United States Department of Agriculture, Division of Forest Pathology, made this discovery, not the tree surgeons. He tested a number of compounds and found that by adding a small quantity of phenyl-mercury-acetate to the asphalt he could make the wound dressing really fungicidal. So far as the writer is aware, this is the first really fungicidal wound dressing devised for use on trees, and it was developed in the 1930's. A professional plant pathologist can hardly be blamed for looking upon certain aspects of tree care with some suspicion.

The above accounts of a few typical plant diseases should serve to indicate that fungus diseases are common, not unusual; that the fungi which cause these diseases, if simple in structure, may be biologically complex; that many of these fungi cannot be controlled easily, directly, or effectively; that often our "control" of the diseases does not involve the elimination of the fungus concerned, but rather is a matter of our keeping a few jumps ahead of a resourceful and tricky enemy. Little has been said about the influence of these diseases on society as a whole, for that is a difficult thing to evaluate. That they have had some profound effects and that they still influence the course of development of communities and societies are obvious to anyone who deals with economic plants.

One modern historical prophet of some fame, in a recent book that was long a best seller, proposed the theory that men have attained their highest development not where life was easy and nature lush, but in the stony soil of adversity. As a case in point, he compared the southeastern United States with portions of New England. In the South, supposedly, life was so easy that man did not have to struggle against nature and therefore did not attain to the heights of the hardy Yankees in their unfriendly environment.

This thesis disregards some pertinent biological evidence. In the South, where to him nature seemed so friendly, she is friendly to many of man's enemies also. Not very long ago terrific epidemics of cholera, typhoid, yellow fever, and malaria raged every few years through the towns south of New England. In the late 1700's Philadelphia was at times very hard hit, and even New York was not exempt. Farther south, nature further manifested her kindness in constant endemics of amoebic and bacillary dysentery, hookworm, and a variety of lesser afflictions. Crop failures in the southern region, from rusts, smuts, blights, mildews, and rots, were so common that some of the staple food crops could not be economically grown, or even grown at all. Insects took their share of plants, as well as of animals, and spread a variety of diseases in both.

Picture, if you will, a poor southern farmer, vintage of 1825, leaning against a sagging wood shack — sagging because its sills and joists are being eaten away by termites and fungus rot. His body is racked with the chills of malaria, his bowels griped and tortured by dysentery, his energy drained by hookworm. Of the ten children born to his parents, he is one of the two who have survived to debilitated manhood. He gazes disconsolately out at his fields, where his cotton plants are being withered by anthracnose and his tobacco cut down by blue mold and rot.

His small wheat-field is heavily rusted, and the seed itself so badly infected with smut that it will not make palatable bread, but he will be glad to get even such bread. He has planted a few potatoes, but late blight has already eliminated most of them. The corn and beans look fairly good, although the beans are beginning to rust and if the weather stays moist they will yield nothing; and the corn already has about 20 per cent cob rot, plus an equal amount of smut. Most of his pigs have trichina, and he and his family will get it from them next winter. This is the land which to the historian in his ivory tower looked so lush, so easy, so free from competition that it did not stimulate man to rise to great heights.

Admittedly the picture is overdrawn somewhat, but it is a more accurate account of the biological factors involved than that given by Mr. Toynbee. The same or similar situations no doubt prevailed in the past in some of the other regions where life supposedly was so easy, as it prevails today in certain portions of the tropics where life seems easy too. This is not to say that diseases in human beings, and in the domestic animals and cultivated plants on which we largely depend, have been the major determining factor in the history of civilizations. But they have been important enough so that any historian or biologist who disregards them can give only an incomplete and distorted account of man's struggle. No serious study of the impact of certain plant diseases on society has been made. If one could evaluate the effect of all these subtle and obvious biological factors on the past and present, it would make most fascinating history. To historian and biologist alike.

Additional Reading

J. C. Arthur, *The Plant Rusts (Uredinales)*. New York: John Wiley & Sons, 1929. 446 pp.

J. C. Boyce, *Forest Pathology*, 2d ed. New York: McGraw-Hill Book Company, 1948. 550 pp.

K. Starr Chester, *The Nature and Prevention of Cereal Rusts as Exemplified in the Leaf Rust of Wheat*. Waltham, Mass.: Chronica Botanica Co., 1946. 269 pp.

C. M. Christensen, E. C. Stakman, and J. J. Christensen, "Variation in Phytopathogenic Fungi." *Annual Review of Microbiology*, 1:61–84 (1947).

E. C. Stakman, "International Problems in Plant Disease Control." *Proceedings of the American Philosophical Society*, 91:95–111 (1947).

E. C. Stakman, "The Nature and Importance of Physiologic Specialization in Phytopathogenic Fungi." *Science*, 105:627–632 (1947).

John Charles Walker, *Plant Pathology*. New York: McGraw-Hill Book Company, 1950. 699 pp.

CHAPTER 7

Fungi in Foods and Building Materials

FUNGI cause heavy losses not only in all our crop plants but also in practically all the goods and materials processed from these plants, and in many kinds of manufactured products derived from things other than plants or plant products. Molds avidly attack and consume or spoil stored seeds, vegetables, dairy products, and meat, decay wood in all its forms and in almost all its normal uses, grow on, in, and through paints and other supposedly protective coatings, rot fabrics and leather, and in general raise expensive havoc with most of the common materials we use. Here again the field is a large one, and the present chapter aims to present only some of the more common and interesting problems posed by these ever present and destructive molds.

Fungi and Stored Grains

It generally is assumed that once a bumper crop of wheat or corn is harvested it is "in the bag" and safe from loss. The grain may be in the bag or bin, but there are many other things in the bag or bin with it, including insects and storage fungi. Most of the insects that infest grain and other stored products are large enough to be easily seen; the damage they cause has long been recognized, their habits have been studied in considerable detail, and control measures have been worked out for most of them. They are still with us, to be sure, but not in the numbers they once were.

The same is not true of the storage fungi. That these might even play a part in spoilage of grain and other materials stored in bulk was first recognized only a couple of decades ago, and only now are we beginning to realize how important a part they actually play, learning something of their habits (and they have some peculiar ones) and accumulating information basic to their control. The following attempts to summarize our present knowledge in the field.

Fungi in seeds. Seeds of cereal grains and of many other crop plants may be invaded by a variety of fungi before the seeds mature. Approximately 50 genera of fungi, for example, have been isolated from seeds of barley, and from a single barley seed it is not at all unusual to culture half a dozen different sorts or species of fungi plus several different kinds of yeasts and a variety of bacteria; a single gram (about 25 kernels) of fairly high grade malting barley — and only the cream of the crop is selected for malting — when ground up and cultured may yield tens of thousands of individual colonies of fungi, hundreds of thousands of colonies of yeasts, and millions of colonies of bacteria, which indicates that seed is not just seed.

The fungi that infect seeds while they still are on the plants in the field we designate as "field fungi." Alternaria, Cladosporium, Fusarium, and Helminthosporium are among the common genera of field fungi, but many others are encountered with greater or less frequency. These field fungi may stain or discolor the seeds they invade, shrivel or blight the seed before harvest, or reduce the germination percentage when the seeds are planted, but they do not deteriorate the seed in storage. All require a high moisture content to grow, about the same amount as do the seeds themselves — in the starchy cereal seeds a moisture content of about 22–25 per cent. Seeds of our common agricultural crops never are stored commercially with moisture contents anywhere near as high as this.

Storage fungi comprise a different group, with different habits. Most of them are in the genus Aspergillus, although Penicillium is met with occasionally too, plus a few others of minor importance. These storage fungi, at least the major ones, in the genus Aspergillus, never invade the seed to any significant degree or extent before harvest, even in years of moist harvest. But if the moisture of the seed is high enough when the grain is stored, or becomes high enough later, these storage fungi invade the seeds and cause various sorts of troubles that result in lowered quality.

If grain such as wheat or corn is stored in bulk with a moisture content of 14–15 per cent, storage fungi invade the germs or embryos of the seeds. The embryo first is weakened, then killed, and later turns black, and kernels with black germs are said to be "damaged" or "sick." The appearance of such damaged or sick seed often is the first indication an elevator man has that all is not well with his grain, and even these dark and decayed germs can be seen only if the pericarps that cover them are removed. For a long time the cause of sick wheat or corn or barley was considered to be a mystery, and there was much speculation as to the possible cause or causes of the condition. Had anyone familiar with fungi bothered to look at the discolored germs closely with a microscope it would have been easily apparent that storage fungi were involved, since spore masses of Aspergillus are common on the surface of such germs, and sometimes even cover and conceal the germs. (See Plates 8 and 9.)

The storage fungi also cause grain to heat, since like all other living things they produce heat as they respire. When insects develop vigorously in a mass of grain they will cause it to heat, too, but only up to a temperature which the given insect can endure — from about 90 to 110 degrees Fahrenheit, depending on the species of insect. However, grain stored in bulk sometimes heats up to a temperature far above this, and may even undergo spontaneous combustion. For a long time there was a misconception about this, also.

It was supposed that, in the absence of insects, the grain itself was respiring rapidly enough to engender considerable heat, although there never was any experimental evidence to support this contention. Work with wheat free of storage fungi proved that the grain itself respired very slowly at moisture contents up to 18 per cent. Even if the grain has enough moisture so that it can germinate — about 25 per cent — and so respire rapidly, it cannot raise the temperature above about 85 degrees Fahrenheit, because the moist grain itself is killed by any temperature above that.

Many practical grain men once believed, and a lot of them still believe, that grain has "an urge to heat and germinate in the spring." That is, about the time that corn and wheat are being planted in the fields the grain in its dark storage bin feels the eternal call of spring, begins to breathe heavily, like an ardent bull, and sweats and heats. Granted that stored grain often heats and spoils in the spring, this romantic explanation of the process is complete nonsense; the same heating occurs in bins of turkey feed, baled cotton and wool, and manure piles, materials that by no possible stretch of imagination could feel the mysterious call of spring, or respond to it if they did feel it.

The answer to the heating problem when insects are absent, and often indeed when they are present but when the temperature goes above that endurable by the insects, was found to be in the storage fungi. Some of these fungi, mainly in the *Aspergillus glaucus* group, will invade grain slowly when the grain has a moisture content of only 14–15 per cent. They do not grow rapidly enough to produce any detectable heating at that moisture content, but they produce moisture as one of their products of growth, and so increase the moisture content of the grain. Once the moisture content is increased to 15–16 per cent, *Aspergillus candidus* and *A. ochraceus* take over. They grow more rapidly, cause some rise in temperature, and release more water. Once the moisture content of the grain reaches about 18 per cent, *A. candidus* grows very rapidly, and is joined by *A. flavus*, which also grows rapidly. These two species like it hot and humid, and *A. candidus* can raise the temperature up to about 130 degrees Fahrenheit and hold it there for weeks, by which time the grain is a hot and stinking mess.

At that point some of the ever present thermophilic or heat-loving bacteria take over, and, given conditions for rapid growth, they can raise the temperature to about 170 degrees Fahrenheit. These bacteria cannot survive temperatures much higher than this, but they produce materials that can undergo purely chemical oxidation, and this may carry the temperature up to the point of spontaneous combustion.

This course of events in heating of materials such as stored grains has been worked out rather precisely and convincingly in the laboratory, and there no longer is much room for doubt that the account given above applies equally well to stored grains, baled cotton and wool, compost heaps and manure piles. Such heating occurs in grain bins at any time of the year, whenever the moisture content and the temperature of the grain are high enough to permit the storage fungi to grow. It is especially common in bins of corn in the spring because then the grain is warmed up enough to permit the fungi to grow rapidly.

Moisture content and its measurement. The really important factor in determining whether storage fungi will invade a given bulk of stored grain is the moisture content of the grain. The lower limit of moisture content that will permit the most drouth resistant of the storage fungi to invade grains such as wheat, barley, and corn is about 13.2 per cent. It would seem simple enough to make sure that grain, when stored, has a moisture content below this, and so to keep the grain sound and in good condition for years, but it is not quite so simple as it seems. Several of the complicating factors are worth a brief mention.

First, moisture content may not be accurately measured. In practice moisture content of grain is determined by means of a machine that measures electrical conductivity or capacitance of a sample of grain. The really accurate method is by oven drying a sample of grain, but this takes too much time to be usable in an elevator or terminal where many samples must be processed quickly. In our experience with several makes of these machines in common use in the grain trade, any one of them may be off by as much as 1 per cent on a given sample when compared with the results of oven drying of the same sample. In the really critical range of moisture content, between 13 and 15 per cent, the machine in most common use frequently has indicated a moisture content from 1 to 2 per cent below that indicated by oven drying, and this was especially true of samples that had begun to be invaded by storage fungi and that were in the first stages of deterioration. Even if the machine is accurate, the moisture content figure it gives on a pint sample of grain does not necessarily indicate the range in moisture content that might prevail in the bulk of 40,000 bushels or 100,000 bushels from which the pint sample was taken.

Second, even though a given bulk of grain may be stored with a uniform and supposedly safe moisture content, the moisture may become unevenly distributed with time. Air circulates slowly through a bulk or mass of stored grain, and if there are fairly large differences in temperature between different portions of the bulk, moisture will be transferred slowly from the warmer to the cooler portions. Such temperature differences are likely to be fairly large in the winter, when the grain near the walls and at the exposed top of the bulk has a lower temperature than that in the interior, and so the cooler portions gradually accumulate moisture. The temperature may be too low for storage fungi to grow appreciably in this moist grain during the winter, but when spring comes and the air temperature rises, this grain warms up, the fungi grow vigorously, and so the grain has an urge to heat and spoil in the spring.

Numbered bags of grain were once buried, each bag containing about a pound of grain with a moisture content of 11 per cent, in the grain of a silo-type terminal elevator in Richmond, Va. The grain going into the elevator had an *average* moisture content of 13 per cent, and so presumably was safe for storage. After three months the grain began to heat, and so the bin was emptied, at which time we recovered our sample bags, and immediately determined the moisture content of the grain in each one. The average of all the 36 bags was 13 per cent, identical with the average for the entire bin, but the grain in some of the bags had a moisture content of 10–11 per cent, and that in others had a moisture content of 17–18 per cent. As determined by the fungi present in the grain in the sample bags (the grain in these samples was almost free of fungi when the bin was filled) all of them at one time or another during the three months had had a moisture content of over 15 per cent for a long enough time to permit heavy invasion by storage fungi. There had been very stormy weather in this bin, and evidently stormy weather prevails in many large bins of grain, contrary to what the grain men choose to believe.

Third, if a bulk of grain comes into a given elevator with a moisture content of 15–16 per cent, too high for safe storage, it is a common practice to mix this with a sufficient amount of grain of lower moisture content to bring the average down below the supposedly critical figure of 14 per cent, which the federal regulations state to be acceptable for all normal grades of wheat and barley. In theory, the moisture content should soon equalize in the mixed grain, and all should be well. There is an abundance of evidence, however, that the moisture content never does equalize completely; that portion of the grain which was originally of higher moisture content will retain a somewhat higher moisture content than the average figure, and the portion originally at the lower moisture content will retain a somewhat lower moisture content than the average. And more than this, which is bad enough in itself, that portion of the grain originally with a moisture content of 15–16 per cent may have had that high a moisture content for several weeks or more. In that time it would have become moderately to heavily invaded by storage fungi, and some of it might be on the verge of deterioration, although this is not apparent to the elevator operator. After being mixed and stored, the deterioration continues, sick or damaged grain develops, with possible heavy financial loss.

Fourth, insects and storage fungi developing in stored grain may increase the moisture content appreciably, not only in that portion of the grain in which they are active, but in a portion of equal or greater volume above this. In some of our tests, weevils introduced to grain with a moisture content of 12–13 per cent soon increased the moisture content to 14–15 per cent, by which time the fungi began growing vigorously too. Within a few weeks more the moisture content had increased to 18 per cent, and after some months more to over 20 per cent.

All this adds up to deterioration occurring in many bins of stored grain that elevator men think should be safe for storage. Judged by the moisture content figures shown on their record books the grain should have been o.k., and so spoilage has been attributed to mysterious causes. It would be easy to take samples occasionally from different portions of the grain after it has been in storage for a time, and determine moisture content, just to see if all is well, but relatively few elevator superintendents do even this. Some practical men are more practical than others.

Detection of storage fungi and predicting storability of grain. It commonly is said that the early stages of deterioration are impossible to detect, and that the man in charge of stored grain has no way of knowing whether he is loading his bin with really sound grain or with grain already in the first but hidden stages of deterioration. It is quite true that one cannot detect the early stages of deterioration by ordinary visual inspection. But if samples are taken from known places in a bulk of grain just after the grain has been put in the bin, and periodically thereafter, tested accurately for moisture content, examined microscopically by someone familiar with sound and deteriorating grain and with storage fungi, and cultured on agar to detect the numbers and kinds of storage fungi present, the present condition of the grain can be evaluated and its storability or deterioration risk predicted with surprising accuracy. Some of the larger grain firms in the United States are doing this very thing, and their losses have been reduced to almost nil. Others prefer to believe that "grain has an urge to heat and germinate in the spring."

Fungi in Flour and Bread

There is no such thing as mold-free flour, but the flour from some mills is consistently more moldy, has a far higher number of fungus spores per unit of weight, than that from other mills. In tests of some 500 samples of flour from many different mills, some years ago, the number of fungus spores ranged from a few hundred to a few thousand per gram of flour, or up to several million per pound. Nearly all these fungi were typical storage molds, primarily in the *Aspergillus glaucus* group. At first it was thought that these spores might be coming from moldy grain, but it soon was found that this was not their source. After some detective work the source or sources were found to be within the mills themselves. Wheat is conditioned to a moisture content of about 16 per cent just before it is milled; as the grain passes through the rapidly revolving rolls that grind it into flour, some heat is produced and some of the moisture in the grain is released. If the roll housings and spout walls are cooler than the air within them, moisture condenses on the inner walls. Flour adheres, and the fungi grow in this layer of moist flour and shower their spores into the passing stock.

A few batches of flour have been encountered in which the population of Aspergillus was heavy enough to give the flour a musty odor and make it unusable for baking, but such heavy contamination evidently is rather rare. In the numbers ordinarily encountered in flour these fungus spores presumably have no significant effect on flour quality.

The heat of baking kills all fungus spores that may be in flour or other ingredients of the baked goods, but bread and other bakery products are subject to contamination from airborne fungus spores as soon as they cool. If the flour used by the bakery has a high population of fungi, the bread almost inevitably becomes inoculated with a large number of spores before it leaves the bakery. Even if the flour itself has a very low count of fungus spores the bakery may supply a heavy inoculum from moldy materials on the premises, and so in a modern bakery good housekeeping is of the essence.

Some years ago a survey by men in the baking industry indicated that they and consumers in the United States were losing at least several millions dollars a year from bread that got moldy before it was eaten. Part of the answer to this was better housekeeping, and the good bakeries now maintain a degree of cleanliness that is not approached in the average home. Calcium propionate or other mold inhibitors are added to bread to keep it free of molds for a few days, by which time it supposedly is eaten. Bread still gets moldy, of course, but usually because a housekeeper may maintain in the breadbox what amounts to a forcing chamber for molds.

Fungi in Wood and Wood Products

Several thousand species of fungi live almost solely and exclusively on wood; wood is their normal and natural food, and as these fungi consume the wood they cause decay or rot. All wood is subject to decay, although different woods differ greatly in the rate at which they will be attacked by wood-rotting fungi under conditions favorable to the growth of the fungi. As a general rule the paler colored woods, such as aspen or basswood among the hardwoods or the true firs among the conifers, will decay much more rapidly than the darker woods, such as black locust or white oak among the hardwoods, or western red cedar or redwood among the conifers. The dark color and decay resistance are contributed not by the wood substance itself, but by materials deposited in the wood as heartwood is formed in the growing tree. Sapwood of all species of trees is comparatively susceptible to decay, while the heartwood of some species is exceedingly resistant. Other things being equal, the denser, heavier woods are likely to be more resistant to decay than the lighter, more porous ones. Under conditions that favor most rapid development of wood-rotting fungi, a fence post or building support of decay-susceptible wood may be almost completely consumed in a year, while the most decay-resistant woods under the same circumstances might endure for a couple of decades.

Conditions necessary for the growth of wood-rotting fungi. Wood-rotting fungi, like most other fungi, need food (the wood in which they are growing), water, some but not very much air or oxygen, and a favorable temperature. Under most circumstances where wood is used the critical thing is moisture content — they require free water, liquid water. The "fiber saturation point" in wood is that moisture content at which the cell walls of the wood are saturated with water, and any additional water taken up by the wood occurs as free water in cell cavities. In most woods the fiber saturation point is at a moisture content of 25–28 per cent, and this is the lowest moisture content at which decay will occur in wood. If, on the other hand, wood is completely saturated with water, so that no free air is available, the wood-rotting fungi are unable to grow, which is why logs submerged for decades in rivers still are sound. Wood submerged in seawater will not decay, but will be attacked by marine borers and chewers, which eat it up even faster than wood-rotting fungi can.

So wherever wood is wet, but not submerged, it will be attacked by wood-rotting fungi. And it need not be continually wet; intermittent wetting such as occurs when summer rains fall on a leaky roof and soak the rafters beneath is sufficient to allow fairly rapid decay, especially if the wood is relatively susceptible to rot, and most woods now in common use are susceptible to rapid decay.

Decay hazards in homes and other buildings. Since most of us live in houses constructed mainly of wood, it may be of interest to look into some of the common decay hazards that such structures are subject to. These hazards will vary some, of course, with the climate, especially with rainfall and relative humidity, with temperature, with the type of construction, and perhaps to some extent with the nature and habits of the people. As stated above, in general whenever wood is wet enough long enough it will decay. How does it get wet enough?

At times green wood, unseasoned wood, is used in construction, wood which has a moisture content of 50–100 per cent or more, based on the oven-dry weight of the wood. If such wood is used in the framing members of a house, such as joists, studdings, and rafters, or for sheathing of walls or roof, it might require some months to dry out, especially in humid weather. During this time decay can become established, and while it is not likely to develop far enough in a few months to cause serious loss in strength of the wood, it is there. Some fungi that cause decay of wood can endure drying for years and revive again as soon as the wood becomes wet. So if such infected wood becomes moist again, even intermittently, in the next few years, the decay may progress. (See Plate 10.)

Houses in the process of construction frequently are exposed to rain or snow that may thoroughly soak much of the exposed wood, and so give an opportunity for decay to get underway. Again this may not be a serious matter unless the wood later is exposed to constant or periodic wetting from rains or other sources. With hard and prolonged rains water will seep into any open cracks or joints, such as those around windows and doors. Once the wood is well soaked it may require a long time to dry out, especially if it is kept well covered with high-quality paint that retards the outward passage of water vapor, and during this time the wood-rotting fungi can work. It is this course of events that leads to decay around windows and doors, where entry of water is somewhat difficult to prevent. With roofs of shallow pitch, snow may accumulate in winter, melt next to the roof, seep down under the layer of snow, freeze near the edge of the roof, and form a mass of ice that causes water to back up under the shingles and into the sheathing, rafters, and studdings below. Such watersoaked wood may require most of the summer to dry out again, and during this time the wood-rotting fungi develop vigorously. After half a dozen or so such annual cycles the wood may be thoroughly decayed, although this is not evident to the owner until the ceiling or wall begins to sag or big gobs of mycelium or fruit bodies of wood-rotting fungi suddenly sprout from the living room wall.

Water can come from the inside, too. Sealing around bathtubs and shower stalls may be imperfect when installed or may develop leaks later, and permit slow seepage of water into the wall studdings or subflooring and joists. With the high humidity maintained in many of our modern homes such wood, once wet, may never dry out thoroughly, and with time it will decay.

Humidifying systems put a surprising amount of water into

the air of a home; to maintain a relative humidity in the "comfort zone" in a three-bedroom home, for example, in the winter in northern United States, may require evaporation of 25 gallons of water per day — and this water has to escape somewhere. If the house is well built it will have insulation in the walls, with a really good vapor barrier on the inside of the insulation. If the house is not well built it may have no insulation, or insulation without a vapor barrier, so the water vapor that passes out through the walls may condense in the insulation or in the wood sheathing of the outer wall, and during the next summer decay is almost certain to begin.

In summer or in any humid weather cold water pipes and toilet tanks drip quarts of water onto whatever is beneath them, and if this water soaks into wood and keeps it constantly moist, decay is almost inevitable. Architects and builders probably could do with a little more knowledge of the biology of wood-rotting fungi than most of them now have, but it can do the homeowners themselves no harm, and might do them much good, to become familiar with some of the common decay hazards and how to cope with them. The general idea is to keep wood dry; if this is impossible, impregnate it with a good wood preservative.

Dry rot. A brief account of dry rot may be in order, because while no such thing exists in nature, since dry wood never will decay, the idea of dry rot definitely exists as a part of our folklore. This idea probably gained wide acceptance from the fact that decay often occurs in wood which to the unaided eye *appears* to be dry. The decayed wood may even *be* dry when it is detected, but it was not dry when the decay was going on. This sort of decay is exceedingly prevalent in the more humid regions, such as the southern Gulf Coast and lower East Coast of the United States, in England, and throughout northern Europe, but it can and does occur anywhere. At one time it was supposed that only a couple of species of fungi caused this sort of decay — *Merulius lachrymans*, known in Europe as the "house fungus," and *Poria incrassata*, but in recent times many other species have been implicated also.

In regions of high average humidity throughout much of the year most of the wood in houses and other buildings may have a constant moisture content of close to 20 per cent, too low for decay to occur, but not much too low. If such wood becomes wet from any cause, it will stay wet for a long time. Basements and cellars in such climates may be continually moist, and the wood in the ground floor and the lower parts of the walls may also be moist enough for wood-rotting fungi to become established. If wood is left piled up or scattered about in such a moist cellar, wood-rotting fungi will invade it, and from this source mycelium will spread up the wall to the floor joists and house sills. Once the fungus begins to decay these, it produces enough metabolic water to keep the wood wet enough so that decay can continue. The fungus may progress upward through the interior of all the wood members of a house, leaving a thin shell of seemingly sound wood on the outside of wainscoting or doors or studdings and floors. The first indication of such decay may be partial collapse of a floor or wall, and by that time the whole house may be so decayed that it is held up practically by faith and a thin outer layer of wood and paint.

This sort of thing is of very common occurrence in England and northern Europe, and is one potent reason for the preference of stone over wood as a building material in those regions. They don't necessarily *like* their living quarters to be cold and dank; perhaps the wood-rotting fungi have forced them into it. But it can happen even in Minnesota. At the University Biology Station at Itasca Park, we occasionally occupied a cabin with a half basement. This basement had cement walls about eight inches thick, and when the walls were built the wood forms were left on the outside. Wood-rotting fungi or a wood-rotting fungus decayed these, of course, then grew through a crack in the wall and spread out in large attractive fan-shaped masses of mycelium on the inner side of the wall. Various sorts of materials, including bolts of aspen wood, were piled on a ledge against the wall, about three feet from the floor, and the fungus soon invaded these, and from there began growing up the wall. Located in a small room just above was the toilet. The bowl of this had been cracked one fall when someone forgot to drain it before freezeup, and so water seeped into the floor and into the joists beneath. The fungus seemed to sense this, and probably did, since in the quiet air of this cellar there doubtless was a gradient in relative humidity, higher near the moist floor above, and the mycelium grew toward this region of higher humidity. Within about a year the fungus reached the floor, then developed vigorously in the joists and flooring, and within several years had decayed them to the danger point. (Plate 11 shows decay in the cabin wall.)

Wood preservatives and wood preservation. Wood preservation has been big business for well over a century, and was practiced even before it was generally realized that decay of wood was caused by fungi. When the railroad and telegraph lines first began to span great distances, decay in ties and poles became of great concern. Wood was cheap and plentiful then, but the costs of replacement of rotted ties and poles, and the wrecks and service failures resulting from rot, gave a great impetus to wood preservation.

The first preservative extensively used, and one that maintained its predominance for about a hundred years, was coal tar creosote, a distillation product of coal, although a number of other compounds, including mixtures of various toxic salts, were also used. Since the 1930's, solutions of pentachlorophenol in various oils have been more and more widely used for this purpose. As a result of wood preservation, railroad ties now fail from mechanical wear before they fail from decay. Also poles and construction timbers of even decay-susceptible species of wood can be made very resistant to decay by impregnation with a preservative, and many of the structural parts of houses and other buildings, such as window and door frames, sills and supports, are treated with preservatives before they are installed. The do-it-yourselfer can buy solutions of pentachlorophenol or other good preservatives under various trade names, and apply them with a brush to any wood exposed to the weather, and thereby lengthen the service life of the wood by many years. Anyone who uses wood in construction should trouble to learn something about wood decay and its prevention by means of preservatives.

Paints and protective coatings. It still is commonly believed that a surface coat of paint or varnish will preserve wood from decay, and one of the large manufacturers of paints and varnishes used to advertise "Save the Surface and You Save All." A fine advertising slogan, but compare it with this statement in a government bulletin of recent vintage (*Decay and Termite Damage in Houses*, 1948): "Paint is not a preservative. In many cases it will protect wood from intermittent wetting, especially if applied to ends and edges as well as exposed faces and so maintained as to allow the fewest possible cracks at joints. In some other cases, as for example when applied to wood that is not seasoned, *it may favor decay by hindering further drying* [italics mine]. In warm moist climates or in rooms with very moist air, paint itself may mold and become unsightly."

It is not only in green wood that paint will increase the decay hazard, and not only in warm, moist climates that paint may mold. Wherever wood becomes moist, either from long-continued rains, from contact with moist soil or other moist materials, or from condensation or leakage, the paint retards evaporation of water from the wood and promotes or favors decay. Paint has some virtues, but protection of wood from decay is not one of them.

Paint itself is subject to fungus attack, some paints extremely so. In humid summer weather it is not unusual for painted walls, especially in basements or other closed rooms, to sprout a crop of varicolored molds so heavy that the wall itself is hardly visible. This is unsightly to say the least, and if the persons who occupy the rooms where the molds are growing, or even the rooms above, happen to be allergic to the spores of some of these fungi, as many people are, it may be a lot worse than merely unsightly. Many homes have a heavy and constant population of fungus spores in the air, spores that are coming from within the home, not blowing in from outside, and painted surfaces are one of the common sources of these spores.

The problem of fungi in paints and what to do about it has been worked on intensively only during the last 10 years, and primarily by men of a commercial laboratory that manufactures fungicides. They exposed painted wood panels by the thousands in several sections of the country, studied the fungi and bacteria in the exposed paints, and tested a multitude of fungicides as paint protectants. They found that fungi were indeed much more important in paint soiling and paint wear than anyone had realized, and they also came up with fungicides that do an excellent job of protecting paints from invasion by fungi.

Fungi in Fabrics

All plant fibers are subject to attack by fungi, some more so than others. Cotton fiber may be invaded, discolored, and weakened by fungi before it ever is picked from the plants in the field. In the laboratory, strips of cotton duck or canvas, inoculated with some of the fungi that decay cotton fibers, and kept continually moist and warm, may be reduced to zero strength within a couple of weeks. In the Second World War this sort of decay was of major importance in military operations, and especially in the humid tropics where clothing, tents, and other fabrics were moist much of the time. In use, cotton no longer is just pure cotton, but is impregnated with sweat, infiltered with dirt, covered with dust, and so made even more susceptible to invasion by fungi. Ropes and cordage made of plant fibers are likewise subject to decay, especially since once they become wet they may remain wet for a long time. Gunny sacks and grain sacks are no small item of expense to those who use large numbers of them for storage and shipment of various goods. Potatoes, for example, usually are collected in the field in gunny sacks, made of Indian hemp or jute, which is very subject to fungus attack. After the sack has rested on the moist ground for a day or two, the bottom is thoroughly invaded by fungi, and decay has begun. If the sack sits on the ground for a week, or if it is stored in a moist place, with dirt adhering to it, it is likely to decay very rapidly, and in the past such sacks normally could be used only one season.

Owing largely to the impetus given to the study of these problems by the Army Quartermaster Corps during the Second World War, the nature of this decay of fabrics by fungi was soon disclosed, and special preservatives for the various fabrics in various uses were developed. This was not so simple as it might seem. Whatever preservative was used not only had to be toxic to a variety of fungi, some of which can stomach large doses of poison, but had to be compatible with the dyes used to color the cloth, compatible with waterproofing chemicals, resistant to leaching by water and sometimes by dry cleaning chemicals, stable when exposed to sunlight, which rapidly breaks down some preservatives, not inactivated by dirt or sweat, and not irritating to the skin.

These various requirements have led to the development of a number of different preservatives especially adapted to this or that fabric or use. Many cotton fabrics now are impregnated during manufacture with chemicals that greatly inhibit fungus attack, and this increases the useful life of some fabrics many times over.

Fungi in Protein Glues

Until the 1940's most plywoods were fabricated with protein glues of one kind or another, and the so-called "interior" plywoods, designed for use indoors, still are. The chief protein glues now used for this consist mostly of milk casein or soybean flour; frequently the protein material is "extended" — which really means diluted — with rye flour, wood flour, or walnut shell flour. Even some of the resin glues used for exterior plywood and supposedly highly resistant or immune to fungus attack also contain appreciable quantities of rye flour or other extender, and so may not be quite so resistant to fungus attack as they are supposed to be.

A number of bacteria and fungi grow enthusiastically in casein and soybean glues, the fungi at a lower moisture content than the bacteria. So whenever the wood bonded with these materials is exposed to a high humidity or becomes moist, the molds and bacteria slowly digest and weaken the glue, and there are many places indoors where the humidity remains high enough long enough to permit microbial decomposition of these glues.

At one time the leading glue chemists theorized that such protein glues lost their strength, when moist, through a process of purely chemical hydrolysis. They had steeped sample pieces of plywood in tanks of water, which resulted in fairly rapid loss of strength of the glues. They then had the glues tested for fungi and bacteria, but by someone who was not particularly interested in finding out whether bacteria and fungi were or were not developing in the glues, and so cultured the samples on agar media in which few or none of the teeming microflora present were likely to grow. Few colonies of bacteria or fungi appeared in the cultures, and so it was concluded that no microflora were growing in the glues; the water in which the specimens had been soaked sometimes stank to heaven, the stink of course coming from protein breakdown products resulting from bacterial decay, but that was disregarded and the hydrolysis theory was hatched. The bacteria and fungi were excluded from the results without being excluded from the tests — a hazardous research procedure that is by no means unique to glue chemists.

Later it was found that when no fungi or bacteria were allowed to grow in such glue it retained its strength even with intermittent or constant and prolonged wetting. Test specimens bonded with casein and soybean glues were put in sterile water, and they weakened only slowly, from bacteria that survived in the glue. Others were put in water to which bacteria isolated from glue were added, and these specimens came apart in a month or six weeks. Still others were put in water to which a good fungicide and bactericide had been added, and these retained their strength for months. Eventually fungicides were developed that could be added to the protein glues without deleteriously affecting the working life of the glue — the time it remains spreadable after being mixed — or affecting other important properties, but which would protect it from breakdown by bacteria and fungi when it was moist.

An almost shockingly large amount of some potent fungicides was required to protect the glues from microbial action. Some of these compounds were so toxic that two or three parts per million in agar would keep bacteria or fungi from growing in the agar, but in glue a concentration of 5 per cent, or 50,000 parts per million, was required to give good protection. It once was a very common practice, in the study of fungicides, to test them in agar, partly because this was fast and easy. If you are looking for a fungicide to protect agar from invasion by fungi and bacteria, this is a very good approach, but if you are looking for a fungicide that will protect protein glue from microbial breakdown, you had better test it in protein glue. If you want a fungicide that will protect wood or paint or cloth or leather from invasion by microbes, the compounds should be tested in these materials, and under conditions that resemble those in which the microbes concerned do their dirty work.

Additional Reading

Edward Abrams, *Microbiological Deterioration of Organic Materials: Its Prevention and Methods of Test.* National Bureau of Standards Miscellaneous Publication 188. 1948. 41 pp.

J. T. Blake, D. W. Kitchin, and O. S. Pratt, "The Microbiological Deterioration of Rubber Insulation." *Transactions of the American Institute of Electrical Engineering.* 1953.

Clyde M. Christensen, "Deterioration of Stored Grains by Fungi." *Botanical Review,* 23:108–134 (1957).

Richard T. Ross, "Microbiology of Paint Films, VII, the Problem and a Solution." *Spotlights* (Official Journal of the Painting and Decorating Contractors of America), January 1959.

G. Semeniuk and J. C. Gilman, "Relation of Molds to the Deterioration of Corn in Storage, a Review." Iowa Academy of Science, 51:256–280 (1944).

Savel B. Silverborg, *Wood Decay in Houses — Cause, Control, Prevention, Repair*. State University College of Forestry, Syracuse, N.Y. Vol. XXIV, No. 1, 1951.

United States Department of Agriculture, *Decay and Termite Damage in Houses*. Farmers' Bulletin 1993. Washington, D.C., 1948.

CHAPTER 8

Fungus Parasites of Microorganisms and Insects

ALL kinds of animals, as well as all kinds of plants and the produce from them, are subject to attack by fungi. Probably few individuals of any species of animals, small or large, sedentary or active, stupid or intelligent, entirely escape infection by fungi at one time or another. Most of these infections are relatively minor and have little or no effect upon the host. Some, however, cause disablement, and a few are fatal, both to man and to the worms that eventually inherit his carcass. The importance of some of these fungi lies in the fact that they help keep the population of certain insects in check and so are a factor in the biological balance. A few have only biological interest and are of no practical significance whatsoever. In the present chapter some of those that parasitize the smaller forms of animal life will be described.

Fungi That Trap Nematodes

Nematodes are small but exceedingly numerous members of the biological community in which we live. They often are called "eel worms," because they look like miniature eels. Most nematodes are less than a millimeter — 1/25 inch — in length, although a few that live in the vitals of insects may attain a length of more than a foot. They are transparent and relatively simple in structure, and to a layman appear to consist chiefly of a digestive tract

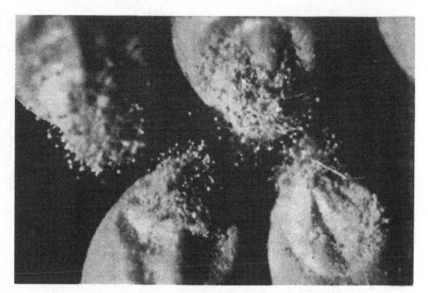

Plate 8. Aspergillus growing from the germs or embryos of grain that has been spoiled in storage by the fungus.

Plate 9. *Aspergillus restrictus*, one of the most troublesome of the common storage fungi, is growing out from the snout and from the anal opening of a granary weevil.

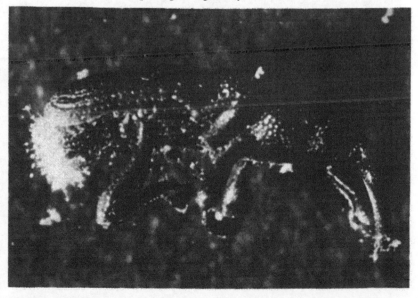

Plate 10. Mycelium of a wood-rotting fungus growing vigorously on supporting timbers in a basement. Photo by William Merrill.

Plate 11. Decay in the wall of one of the faculty cabins at the University Biology Station, Itasca Park, Minnesota. Photo by Dr. D. W. French.

surrounded by muscle, with a mouth at one end, a tapering tail at the other, and a bare minimum of essential organs in between. No fancy or decorative gadgets grace their chassis. Their aims are to eat and to reproduce, and they do both constantly and almost simultaneously. Organisms that live in the soil, where the competition is fierce, are likely to concentrate on these two activities.

Most soil is literally filled with nematodes. Some of them invade the roots of plants and cause diseases, especially in tropical and subtropical countries, and for some reason the study of these nematode-caused diseases has fallen within the province of plant pathologists. In the United States such nematode-caused diseases take a fairly heavy toll of certain crops at least as far north as the Central States, while in subtropical regions such as Hawaii, nematodes constitute a major hazard in the growing of certain commercial crops. Some nematodes invade insects and kill them. Hordes of nematodes live saprophytically in the soil, where they subsist on the partly decayed remains of plants and animals, and probably feed also on bacteria, molds, protozoa, and other small forms of life. So much for the general habits of these small but exceedingly numerous and active fauna of the soil.

The nematodes which live a free and independent life in the soil encounter various hazards. One of these is certain fungi that have found these animals to be choice food; so much so that certain molds subsist almost entirely on a diet of living nematodes. This requires some ingenuity on the part of the fungus, because nematodes, while small by our standards, are hundreds of times the size of the molds that trap them. Also they are strong, slippery, elusive, and constantly moving — seemingly difficult game for a stationary and delicate fungus to trap and invade. Certain molds, however, have devised some ingenious ways of catching these worms on the squirm, and have become surprisingly adept at this sort of big-game hunting.

Let us observe a typical one of these predacious killer molds. The fungus grows through the soil with ordinary branched mycelium, like the mycelium of a thousand other fungi. At frequent intervals short branches grow out from this mycelium, and these join to make minute loops or snares. These in turn may form additional snares, as shown in Figure 7. There are likely to be thousands of nematodes and yards of mycelium of this predatory mold in any teaspoonful of ordinary black soil in your garden or even in the flower pot on the window sill. In his restless, sightless,

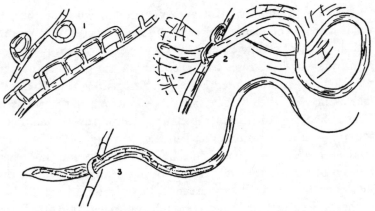

Figure 7. Nematode-trapping fungi. (1) Two types of snares formed by different species of fungi predacious on nematodes. (2) A nematode has poked its head through a snare, the snare has tightened, and the nematode is squirming in its struggle to free itself. (3) The nematode has been killed, and the fungus has grown into its interior.

constant search for food, a nematode will occasionally, by ill chance for him and good chance for the mold, stick his slender, tapering head through one of these many small loops on the mycelium. Contact with his body causes this living snare to contract quickly and powerfully, and the worm is caught. He struggles violently but briefly; the delicate mycelium, however, is flexible enough to yield with his threshing, and the strangle hold seldom is broken. Occasionally the worm will twist about enough to break the snare loose from the parent mycelium and will crawl off with the snare still about his neck. But whether he does or does not break the snare, he is strangled into submission within an hour or less, and shortly thereafter a branch from the snare grows through his body wall and into his nutritious and still living insides. There it grows vigorously and soon digests the worm from the inside out, completely, except for the outer body wall. The food so obtained is used by the fungus to produce more snares to catch more nematodes in, and so on.

There are many different kinds of these nematode-eating fungi, and not all of them follow the same pattern. Some secure their prey by means of a constricting snare, as described above, others by a snare with adhesive material on the inner side, others by a snare that snaps into a figure eight when the unwary worm pokes his neck into it. Some have just single snares in a row along the mycelium, others a complex lattice work with many openings, each of which can entrap a worm.

These predacious fungi are common in almost any soil, and their abundance is evidence that, from their standpoint, this fashion of getting a living is an excellent one. The nematodes eat lesser organisms, are in turn caught and digested by fungi, and the fungi doubtless are in turn consumed by other forms of life. All contribute to the making of the good earth from which we get our sustenance. These multitudes of different kinds of organisms that live in so many different and often implausible ways in the soil, that compete so viciously, and that come to such bizarre ends, are a vital part of the topsoil we are now trying to conserve, and hence an essential part of our basic wealth. They *are*, in fact, the soil. Soil is alive, not dead. We know discouragingly little about what goes on in it, not because soils men or biologists do not think the study is unimportant, for many of them do, but because it is such a difficult field that relatively few have gone into it.

It should be obvious that some of these nematode-hunting fungi are specialists of a rather advanced sort. Certainly millions of years must have been required for their evolution. Delicate and clever adaptations such as they have do not arise suddenly and out of the blue. Both the worm and the fungus that entraps it can be cultivated readily on agar in dishes, so that the whole process can be watched under the microscope. In some of those we have observed from time to time, such large numbers of nematodes have been caught on clusters of snares along the mycelium that it seems unlikely chance alone could account for so many heads being stuck into so many snares. Though we have no proof, it looks as though the fungus may produce some substance attractive to the victims. Man was fairly well along on the road to civilization before he had the wit to make snares, and no man-made snares can equal those designed by the nematode-snaring fungi. It is another example in support of the argument advanced previously in this book, to the effect that although fungi may be simple in structure, some of them are marvelously adapted to the environment in which they live, and many of their complexities we can at present only describe, not explain. Certainly one may legitimately doubt whether nematodes, and the molds which catch and consume them, were created at nine o'clock in the morning on September 9, 4004 B.C. — the hour that Archbishop Ussher in the mid-seventeenth century named as the time of creation.

Some of the relatives of these molds that catch nematodes prefer a diet of protozoa or of other kinds of the microscopic animals with which our soil teems. Each fungus has its own favorite game, and its own methods of stalking and securing it. Most of them are specialists, and it is partly by virtue of their specialized abilities that they are able to compete in the constant and vicious struggle that goes on in the soil. To these organisms the term biological balance is not just a term, it is life and death. Many of these rather remarkable predatory molds are so common in almost any soil that anyone who has a microscope available and who is willing to put in a few hours learning how to make an agar medium — less complicated by far than learning how to bake a cake — can grow both the prey and the fungus predator in quantity. Yet most biology students, whether in grade school, high school, or college, are more familiar with dinosaurs than they are with the interesting things like this that they track into the schoolroom on their shoes.

Fungus Parasites of Insects

In Chapter 4 you read about some of the insects that have learned to cultivate fungi for food. In the biological game as well as elsewhere, the tables can be turned. Whether the ambrosia beetles or leaf-cutting ants are subject to diseases caused by fungi we do not know, but considering the environment in which they live it would be remarkable if they were not. Many insects are.

It is well to bear in mind that not all insects are injurious. The honeybee and the silkworm are economically important and beneficial insects, and both are subject to fungus diseases. Many insects, however, are noxious, and nothing but a nuisance to man. As fungi which attack weed plants may be beneficial, so from our narrow human standpoint some of the fungi that attack such insect pests are beneficial too. Not a great deal of work has been done with fungus parasites of insects, but we do know something about some of them.

Empusa disease. The common housefly, which commutes with such enthusiasm between the privy and the pantry, or, in more elegant surroundings, from the excrement of the pedigreed dog on the neatly clipped lawn to the food on the fancy wrought-iron table on the terrace, often is attacked by a fungus. Unfortunately this fungus, *Empusa muscae* by name, is not a good enough parasite to be of much aid in reducing the fly population, but it still is interesting. (See Plate 12.)

Attacks by *Empusa muscae* probably occur to some extent throughout the year, but they usually are more noticeable in the fall, when the flies become sluggish. The fungus sometimes causes mild epidemics not only in the housefly but in various of his wilder relatives. Invaded by it, the fly no longer is happy and carefree. He walks slowly across the window or wall, his feet spread rather far apart for balance, each one raised and put down carefully and with exaggerated precision, like a stage drunkard. He advances erratically, by fits and starts, with rests for breath and meditation in between. He sometimes buzzes his wings without taking off, and when he does become air-borne, seems to be trying just short experimental flights, as if he were testing a new model, and some of the flights are amateurish. Soon he grounds himself, and well he might, because his innards are being rapidly digested by the parasite. Soon after the outward symptoms of inward distress have become apparent, he expires. If he is standing on a windowpane when death overtakes him, he will, within twenty-four hours, be surrounded by a white halo — a halo made up of sticky spores of the fungus that killed him.

When the fly dies, the fungus that has taken possession sends thousands of short stalks out between the segments of his body. On the tip of each stalk a spherical spore is formed. When mature, this spore is shot off with some force. It is sticky, and so adheres to whatever it lands on. If it lands on another fly, it germinates, grows through the body wall, mycelium develops in the blood, invades the vital organs, digests them within a few days, and so kills the host fly. If the spore lands on the leaf of a tree or on a grass stalk or on the soil, as it so often must, it is not hopelessly lost. It germinates there and produces a short stalk with a spherical, sticky spore at the tip. This secondary spore is somewhat smaller than the first one, as would be expected. It too is shot off. If it lands on a leaf, it will in turn germinate and produce still a third spore, likewise shot off in search of a fly. The fungus cannot go on indefinitely raising itself by its bootstraps in this fashion, because the food from the original spore eventually is all used up. But the device is a rather clever one to give the fungus several shots at the fly. Where flies congregate in large numbers, the fungus may cause epidemics of some size, but apparently the disease is a relatively minor factor in the control of the population of this insect.

Crickets and grasshoppers are subject to attack by other species of *Empusa*. Whether fungus disease is one of the diseases that help to reduce plagues of grasshoppers we do not know, but if so, it is not so regular and dependable as we should like. Sometimes, however, when these pests are present in large numbers and the weather is warm and humid, it may become epidemic and kill large numbers of them. Its spread among them is aided to some extent by the habits of the grasshoppers themselves. They are addicted to cannibalism, as many of our wild friends and enemies are. A grasshopper dying of a mold infection is likely to furnish a meal for several of his hungry fellows, most of whom thereby become infected with the same mold, succumb to the infection, and are eaten in turn by their brothers.

Massospora disease. The seventeen-year locust, which spends sixteen years and nine months of its life in the soil, and only about three months as an adult on the surface of the earth, is attacked by a closely related genus of fungus, Massospora. This fungus invades the abdomen of the locust, digests most of the vital organs, and converts them into masses of powdery spores. As the fungus develops and produces its spore masses within the insect, the invaded locust sheds successive segments of its abdomen, thus allowing the spores to escape. Infected locusts are commonly found crawling around with only a head and thorax, and one or two remaining segments of abdomen filled with the spores of the fungus, most of the supposedly vital organs of the insect having been invaded by the fungus, converted into spore masses, and sloughed off as the fungus progressed.

Fungi not very distantly related to those on flies and grasshoppers are parasitic on a number of other insects, and in portions of California and Florida have been given credit for heavily reducing the population of aphids or plant lice that infest and damage citrus fruits. But when one of these fungi was once introduced into eastern Canada in the hope that it would control some of the insects injurious to apple trees, this form of control proved to be no mere matter of transporting fungi from one orchard to another. That fungi parasitic on insects do, however, reduce insect populations was proved once unintentionally in the citrus groves of Florida. Orange trees were sprayed with a copper fungicide to control some of the fungus-caused diseases of leaves and fruit. The fungicides controlled the fungus diseases of the trees well enough, but they also controlled the fungus diseases of the scale insects and so allowed the scales to increase rapidly. A few months after the trees had been sprayed, there were anywhere from twenty to a hundred times as many scale insects on the fruit and leaves of the sprayed trees as were present on the trees to which no fungicide had been applied. We have not been able to alter the biological balance in our favor by spreading the fungus around. We have been able to alter it in the insects' favor by killing, with a fungicide, such fungi as are present. The unexpected results that have sometimes followed when we altered the biological balance convince us that we do not yet have the information or technics necessary to master nature completely.

Cordyceps disease. Another genus of insect-devouring mold has the name of Cordyceps — literally, *club head.* Species of Cordyceps invade principally the larvae of insects. We do not know very much about how they operate, because most of the insect larvae attacked by them live beneath the surface of the ground, and so the course of events is not easy to follow. We do know that once a larva has been invaded by this fungus, it is done for. The fungus grows throughout the body of the victim, and converts it into a mummy consisting largely of dense mycelium. From this storehouse of food, the fungus sends one or several stalks up through the soil and into the air. This stalk usually is brightly colored, and may be from one to several inches tall. The tip of it is a tapering club, embedded in the surface of which are spherical chambers in which spores are produced. From it the spores are shot out forcibly, to be picked up by the wind and eventually infect the next generation of their host. One or more species of Cordyceps occasionally cause minor epidemics among white grubs — the larva of June beetles. These larvae live in the soil and often do fairly heavy damage by gnawing off the roots of young trees, shrubs, and pasture and forage grasses. Whether the fungus has any major effect on the June beetle population we do not know; probably it does not; but at least a few minor epidemics have been described in which a majority of the larvae in a small area were parasitized and killed by it. In such a case a veritable forest of these brightly colored fungus clubs will appear above the surface of the soil. (See Plate 13.)

Another common species of Cordyceps invades the tortoise-shaped scale insects that live on the bark of hardwood trees and suck the juice from the twigs; it produces from one to several miniature clubs that grow out from the surface of the dead scales. Most of the species of Cordyceps so far encountered attack insects that are injurious to our crops, and so are to be encouraged. Attempts to increase such beneficent fungi artificially, however, or to introduce them into new areas as a form of biological warfare have not met with much success.

Laboulbeniales. Some fungi, and the insects they parasitize, are of no apparent importance to mankind whether for good or for ill. Yet these fungus parasites are interesting enough in themselves to justify a brief look at them. Among them are a considerable number of closely related species in the order Laboulbeniales. That long name can be disregarded, the nature of the things is what counts. These fungi are interesting for several reasons. They are widely distributed on various insects, especially beetles, throughout the world, and some of them can be found almost everywhere that certain beetles occur. A number of them grow on water beetles. So far there is nothing particularly unusual about them. But one species of this group of fungi may be found on only a single joint of the left hind leg of a given beetle. It does not invade other joints of that leg, or get in the beetle's mouth whiskers, or on his wing covers, thorax, or chin. It is restricted to that one spot on just one of his legs, and is found there and there only. This is specialization with a vengeance.

Also the fungus in general seems to use the beetle for little more than support — a place on which to grow, an anchorage. It penetrates into or through the very thin layer of chitinous armor with which the beetle is covered, but does not venture into the interior of the creature where it might find a rich source of food. It forms little more than an additional and foreign bristle on an already bristly beetle. From the insect's standpoint the fungus is of no consequence whatever, causes no discomfort, no loss of function, no impairment of such activities as beetles are addicted to. There are a few exceptions to this, but they can be disregarded for the present.

Further, the fungi in this order, the Laboulbeniales, are off in a group by themselves, and none who have studied them have the slightest idea of what they have evolved from or where they are heading, in an evolutionary way. They have some features in common with certain red algae; some students have taken this to mean that they have evolved from the red algae or from the same basic stock; but this is pure speculation. It could equally well be that because they are living on beetles in water, similar to the environment in which red algae live, they have evolved similar structures. We don't know, and we never shall, because there is no way to get reliable, clinching evidence.

Not only are these fungi mostly of little or no significance to the beetles on which they grow; most of the beetles infected by them, as has already been said, are of little or no practical significance to us. This may not be true of certain ground beetles that are avid hunters of other insects, but it is true of some of the water beetles on which these fungi have been chiefly studied. An inconspicuous and almost invisible fungus growing as just a minor tuft of bristles on an inconspicuous and unimportant beetle. Yet Dr. Roland Thaxter put in several decades at Harvard studying these peculiar fungi, and published several thorough, beautiful, and expensive monographs on them, thereby winning a moderate quota of fame in the mycological world.

Is this type of research valueless? The answer, of course, will depend upon what a person thinks "value" is. Because many aspects of research have been exploited commercially, and have served to make money for stockholders, people in general are inclined to judge scientific contributions by their cash value, even as they judge books or paintings. As our society is organized, most research must have a practical slant, at least for those of us in experiment stations, but over a long period the civilizing aspects of research, not subject to definite measure, and definitely not assayable in cash, may turn out to be one of the most valuable contributions of research to our way of life. First-class research is as rare, and as valuable to the mind and spirit, as first-class music, sculpture, or literature. Its justification is the same as theirs, that by enlarging and feeding the intellect, it ennobles and enriches human life.

Dr. Thaxter's research on Laboulbeniales was of this kind. This is not to say that it did not have some practical value too. In his laboratory men were trained to investigate fungi — to investigate them thoroughly, intelligently, and imaginatively. Some of these men have since contributed a great deal to the practical solution of various problems in industry and agriculture. Dr. Thaxter might have been just as inspiring a teacher if his own interests had lain in a practical direction, but that is doubtful. Fundamental research, which has as its aim only the discovery of more and more of nature's secrets, seems necessary to provide the environment — the backlog of information and ideas — that spawns usable discoveries. "God bless the higher mathematics," as a mathematics professor is reported to have said, "and may they never be of any use to anybody." One may say the same of mycology.

Biological Warfare against Insects

Since the idea of biological warfare against insects has been seriously tested at various times and in various places, it deserves a short summary. We may perhaps think of biological warfare as a strictly modern concept, one of the unpleasant products of our own struggle for survival. But one aspect of it, the possibility that fungi parasitic on insects might be used to control insect pests, was at least suggested, if not actually tried, in the 1830's, some time before most biologists even realized that fungi were living plants. Fairly extensive work by Russian scientists in the 1880's suggested that this sort of control would be worth looking into more thoroughly, and, since then, workers in many countries have attempted to control various injurious insects by spreading around specific mold or bacterial enemies of them. It is a most attractive idea. If practicable, it would mean that instead of waging a constant, expensive, and never-ending battle against some of these six-legged enemies that eat up so much of our crops, produce, and goods, we could merely inoculate them with a fungus, and let the fungus take over the job of destroying them. But the *if* has turned out to be a big one. As the problem has been gone into more thoroughly from various angles, the early hopes of easy control by this means have been greatly modified. Or, if you like it bluntly, most of such attempts have been partial or complete flops.

So far as fungus diseases of insects are concerned, the main facts are about as follows. More than fifty fungus diseases of insects are known. A number of these appear in epidemic form at times, and destroy their insect hosts in large numbers. Yet, when the parasitic molds have been grown and increased in the laboratory and then sprayed on the insects, the percentage of kill seldom has been greatly increased over that which occurred naturally. In no case has an insect pest been totally eliminated by this means, and in few cases has the population of injurious insects been reduced enough to make up for the cost of the operation. Involved in the problem also, though this has not been established, are probably some of the same complexities of variable susceptibility among the hosts, and variable virulence among races of the parasite, that are encountered with the fungus parasites of plants.

The chief difficulty appears to be that before these epidemics of fungus diseases can become established among the insect population, just the right weather conditions have to prevail. Given the right environmental conditions, the parasitic fungi naturally present increase and spread with sufficient rapidity so that any further addition by the hand of man does not increase the epidemic measurably. Where a parasitic fungus not previously present is introduced into an insect population, it may be partially effective in holding the insects in check for a time. But a balance soon is struck. It will be a fluctuating balance: when conditions favor unusual increase of the parasitic fungus, more insects will be infected and killed; when conditions are unfavorable, the epidemic will subside and the insects will increase. Most insects have a phenomenal reproductive capacity. We may kill off 90 or even 95 per cent of the population one year, and the 5 or 10 per cent remaining will, within a year or so, have built the population up to its original numbers. No fungus parasite of insects has yet been found that would regularly cause a mortality approaching 90 per cent. A few have been used with some minor success, and occasional circumstances may arise in which this form of biological warfare will be a valuable weapon against the insects, but so far the results have fallen far short of the original hopes.

Since attempts are now being made to develop biological warfare that may be applied against peoples, the work in the same field that has been done with insects is of some general interest. Remember that the insects against which this type of warfare has been used have been, comparatively speaking, entirely defenseless. They have had no public health service, no sanitation corps, no expert Food and Drug Administration, no quarantine and inspection services, no research staffs to develop counter measures, and so on. Fungi have not been the only enemy marshaled against them, nor even one of the more formidable ones. Bacteria, protozoa, predatory and parasitic insects, predatory animals, fish and birds – all have been brought into play. Some of these subsist solely or largely on an insect diet. All in all, some of the insects on which biological warfare has been practiced have some far more fearsome enemies than those that could be turned loose against us. Work in the field has been going on for nearly a century in many countries, and a good deal of research has been done on both the basic and practical aspects. Yet the results have seldom been very terrible from the victims' standpoint. That our crops, domestic animals, and even we ourselves might suffer fairly heavily from the introduction of destructive bacteria, molds, insects, or other enemies is a possibility that we must recognize. But that we are likely to be wiped from the face of the earth by such means is certainly a very remote possibility.

Remember also that till a few decades ago, the exchange of evil molds, bacteria, insects, viruses, and other agents of destruction went on between nations with scarcely a hindrance. This exchange of vicious parasites has caused a heavy monetary loss, but the potato blight in Ireland is almost the only case of actual national calamity resulting from the introduction of a foreign fungus. Quarantines have not eliminated the spread of human diseases from one country to another, or of plant and animal diseases either; they have merely slowed it down, and we are still exchanging parasites unintentionally at a moderately active rate. Yet relatively few of those introduced become major pests. We need not be complacent, to be sure. Neither, however, is there a biological basis for the type of hysteria that the idea of biological warfare or germ warfare has been greeted with in some sections.

Additional Reading

E. A. Bessey, *Morphology and Taxonomy of Fungi*, pp. 172–183. Philadelphia: Blakiston Company, 1950. 791 pp.

C. E. Burnside, *Fungus Diseases of the Honey Bee*. United States Department of Agriculture Technical Bulletin 149. 1930.

Charles Drechsler, "Several Additional Phycomycetes Subsisting on Nematodes and Amoebae." *Mycologia*, 37:1–31 (1945).

H. L. Sweetman, *The Biological Control of Insects*. Ithaca, N.Y.: Comstock Publishing Company, 1936. 461 pp.

CHAPTER 9

Fungus Parasites of Fish, Land Animals, and Man

THE general field of fungus parasites of animals has not been so carefully explored as the field of fungus parasites of plants. Part of the reason for this is that throughout most of the world the diseases of man and of the lesser animals caused by fungi are fewer in number and of less practical importance than the diseases caused by other organisms and infective agents. For the rest, fungus diseases of animals have not always been recognized for what they are; even where important and destructive, they have often been overlooked or confused with other things. A person sees what he is trained to see, and discovers more easily what he is looking for than what he is not, and for a long time no one was looking for fungi. Some of the fungi that invade the lesser animals are more obvious and easier to recognize than those that invade the higher, and they are the ones that have been studied the most thoroughly. A few of the diseases about which we have learned something will be taken up in this chapter.

Fungus Diseases of Fish

Certain fungi may be a factor of some importance in the rise and fall in the population of various kinds of fish and game. That a host of fungi are present in fresh-water streams and lakes, as well as in the ocean, we know; but we know relatively little about them.* It is a reasonably safe guess that they are at least as important in the life and survival and general distribution of aquatic plants and animals as they are in the life of terrestrial plants and animals. But relatively few studies of them have been made. Up to now, aquatic biology, like soil biology, has largely been in the province of what may be called descriptive science. These descriptions tell little about how the animals live. It is as if a man from Mars had come down to earth and caught up a few men, taken them back, pinned them onto boards, and from the specimens described mankind. He would not know very much about how we operate. We have concentrated on the obvious, and perhaps justly so. Now we are beginning to explore some of the biological mysteries of soil and water.

One genus of water molds, Saprolegnia by name, is so common that it can be recovered from almost every teaspoonful of water from lakes, streams, ponds, puddles, and even birdbaths. One species of this fungus, *Saprolegnia parasitica*, sometimes invades the skin of fish and other aquatic animals, consumes their scales, skin and flesh, and kills them. Thomas Huxley, the great comparative anatomist of the nineteenth century and effective champion of Darwin's theory of organic evolution, apparently came to minor grief over this fungus. In an often entertaining book, *Advance of the Fungi*,† Large has a good account of Huxley's part in this small scientific fiasco, and since Large wrote, more has been added to the story, as still more will be added in the future.

It seems that in the 1880's salmon in the rivers of England were dying in tremendous numbers, and depriving the flower of English knighthood of some of their outdoor sport. Huxley was appointed by the government to cope with the problem. He soon found that the dying salmon were almost always covered with a heavy coating of fungus. After some detailed work he concluded that he had proved the fungus to be the cause of the disease, and

* There are, it is true, books about aquatic fungi that describe the fungi we find. But we still have no idea about how these many fungi found in the water live, how they influence the multitudes of other life in the water, and what they mean to us. Here is a practically unexplored field.

† E. C. Large, *Advance of the Fungi* (1940).

he made some recommendations for its control. The disease sub-sided somewhat, and Huxley returned to his brilliant work in anatomy for such short time as was allotted to him.

Later research, however, of a more careful sort than Huxley's rather cursory job, seemed to prove that the fungus, *Saprolegnia parasitica*, was only a secondary invader of the fish. The fish were apparently first invaded by a bacterium, which was what did the dirty work. Once the fish were dying of bacterial infection, Saprolegnia took over, and covered them with a heavy coating of fungus. It merely gave the finishing touch to a *fait accompli*; it was not in itself able to cause disease. After this, in mycological circles, Huxley was sneered at as a tyro, himself a fish out of water. This is where Large's story ends.

As so often happens in biology, this latter "proof" was only partial truth. What the later experimenter proved was only that the strain of Saprolegnia he used did not parasitize the fish he tested, under the conditions in which he maintained both the fish and the fungus. This is a long way from proving that *Saprolegnia parasitica* is not parasitic on fish, as will be seen directly.

Not so many years ago a research worker at Harvard studied the problem again. He found that a single collection of *Saprolegnia parasitica* was very parasitic indeed on a number of different kinds of fish. He inoculated the fungus into the water in which the fish were living (in aquaria), and it infected and killed them. Also it was able to invade and kill some near and distant relatives of fish, such as newts, frogs, and eels. Some kinds of fish were invaded and killed by the fungus only after they had been in-jured, but others were invaded and killed without having had bruise or injury. Thus, according to his tests, *Saprolegnia parasitica* was a primary invader and killer of fish and some other aquatic animals. He got a culture of the same fungus from Eng-land and inoculated it into the water where the fish were present. It caused no disease. How could this discrepancy be explained?

There is no discrepancy at all. *Saprolegnia parasitica*, like the stem-rust fungus, and like most other micro- and macroorgan-isms, is not just a uniform species. Probably many different strains or biotypes of it exist. Some of them are parasitic on fish, some are not. The worker who followed Huxley on this problem and seemed to have caught him far off base, may have happened to get a nonparasitic strain of the fungus. Certainly the worker at Harvard had a parasitic strain of it; his tests were so well organized, were so well replicated, and included so large a number of each kind of fish that there can be no doubt that in his tests *Saprolegnia parasitica* invaded and killed the fish. Why doesn't it do so in nature, more regularly than it does?

In nature, fish usually live in lakes and streams, with plenty of fresh water. If they are in rearing ponds, Saprolegnia sometimes kills them off in large numbers. Usually only when the water is fouled with wastes of one kind or another, or the fish are overcrowded. What the man at Harvard proved was that the strain of *Saprolegnia parasitica* he had, invaded and killed some kinds of fish maintained under the artificial conditions unavoidable in aquaria. That was all he claimed to prove. Also that under the same conditions another strain of the fungus did no harm to the fish.

Some strains or biotypes of the fungus will invade and kill fish when the fish are living in an environment not very favorable for them. But under conditions even more unfavorable for the fish other strains of the fungus will not. The appearance of the salmon disease in England coincided with a general pollution of rivers from industrial wastes. Huxley may well have been nearly right in his contention that Saprolegnia was invading and killing the salmon. The investigator who followed him may have been nearly right in maintaining that a bacterium was the main cause of mortality, since in the fish he studied this may have been the case. The investigator at Harvard unquestionably proved that there are both parasitic and nonparasitic strains of Saprolegnia. Those who work with fish hatcheries and rearing ponds have good evidence to claim that Saprolegnia is a serious parasite only when the water becomes so defiled that the fish are weakened.

The chief justification for so detailed an account of Saprolegnia disease of fish is to show that these biological problems seldom have simple answers and that even biological truth is difficult to attain by the experimental method. A single experiment often proves nothing or, what is worse, makes the wrong answer appear to be right. Even repeated experiments, with large numbers of individuals and adequate controls, often tell us only what happened under the particular set of conditions maintained in the experiment, and so disclose only a portion of the truth. The jump from the particular fact to the general conclusion is much more risky in biology than in mathematics. What appears to be just a short and easy spring may land the jumper in a quagmire of qualifications, and the smart research worker becomes cautious about ever leaving solid ground.

Several diseases of fish caused by fungi other than Saprolegnia are known, a few of them of at least moderate importance, and probably many more occur than have yet been described. Other aquatic animals also are subject to invasion by fungi. Crabs and crayfish often are invaded by peculiar molds that grow in tufts from their anal opening. Whether these fungi cause any discomfort, bodily or socially, we do not know — the fungi alone have been described, not their effects on the biology of the animals they inhabit. Neither do we know whether the succulent shrimp and the luscious lobster succumb to fungus diseases, but it would be surprising if they did not. And fun to find out. Recent evidence indicates that the reticent oyster sometimes is invaded by a fungus much like one that invades human beings, and that this fungus causes a disease that may be important in the productivity of certain oyster beds.

Fungus Diseases of Land Animals

Some fungi infect birds. The most common of these, which are the only ones yet known, invade the lungs and air sacs of chickens, turkeys, pigeons, and some of their wild relatives. Sometimes these fungus infections lead only to chronic distress and general malaise; sometimes they are fatal. Wild ducks are probably an ideal soil for some of these fungi, but neither Ducks

Unlimited nor the mycologists are able yet to give us any information on this subject.

Pigeon fanciers sometimes become infected with *Aspergillus fumigatus* from the birds they handle. Here, too, one is faced with biological complexities. *Aspergillus fumigatus* causes a fairly common infection of the lungs of poultry; yet if one were to collect it from a hundred different sources, only a few strains of it would be able to cause the infection. Whether the strain that infects the lungs of birds and the lungs of man is a peculiar and parasitic strain we do not now know. The fungus is generally present on a great variety of agricultural seeds, as well as on straw and similar plant debris, and nearly all of the common strains are harmless, to either men or birds. It is another case of there being many different strains of each species of fungus, but though all of them may look alike, they have vastly different effects.

Rodents are commonly subject to fungus attack. Normally this would be of little interest to anyone but mycologists and the rodents themselves, except for the fact that some of the fungi prevalent in rodents also infect men. Mice, ground squirrels, and some of their kin, throughout much of the western region of the United States, suffer commonly from a fungus infection of their lungs. The same fungus also infects men, sometimes with fatal results. To the minds of most people, probably no more unpromising and wasteful research could be devised than that upon the fungi that inhabit wild mice and ground squirrels; yet such study has been basic to an understanding of the epidemiology of this rather serious disease in human beings. Here is another argument for training men to be good research workers, and then turning them loose to find out what they can find out. Some of them will exploit such freedom for their own advantage. Some of them may even be screwballs. But given just the bare modicum of support and the maximum of freedom, enough of them will come up with basic and practical information to pay the cost of the program ten times over.

Whether any of the fungus diseases of wild animals are involved in the regular rise and fall in the population of these animals nobody knows. They are about the only thing that hasn't been brought in to help explain this periodic fluctuation — sunspots and the positions of the planets included. It would seem likely that the study of diseases of grouse caused by fungi, bacteria, viruses, and other microorganisms might come closer to answering the problem of the fluctuation in the grouse population than trying to correlate — as has in all seriousness been done — this rise and fall with sunspots. The writer does not love to hunt and would just as soon see a coyote as a grouse or a deer, and rather see any one of them on the loose in their natural haunts than see a hunter. But if hunters are seriously concerned about the animals they wish to murder, they could do worse than support some work on the biology of these animals, including studies of their diseases.

Fungus Diseases of Man

Man, like other animals, is subject to attack by a number of different fungi, but up to recent times we have known little about them. Fungi could hardly be recognized as a common cause of disease in human beings until medical mycologists had developed the technics to detect fungi in diseased tissue, isolate them, and prove by inoculation into experimental animals that they caused the diseases with which they were associated. The first fungus disease of man was described about 1880, so that the field of medical mycology is still a relatively new one. The numerous specific texts devoted to the field in the last twenty or so years, however, is at least an indication that fungi are now acknowledged as a factor in human health.

What this influence on the general health is, we shall better be able to evaluate as our knowledge of them increases. In the temperate zone none of these fungus diseases of man are so serious a problem as those caused by bacteria, viruses, and protozoa. In the humid tropics they may be of major importance. They are one of the things, as was found out in the second World War, that those who escape to the wondrous islands in the South

Pacific escape to. But fungus diseases, we gradually are learning, are common, widespread, and occasionally serious in the more northern regions too. Of those that attack the interior of the body, some simulate pneumonia or tuberculosis; others produce symptoms so devious that even the expert has difficulty in diagnosing the source as a fungus infection.

Fungus infections of man seldom are fatal. Not infrequently, though, they are disabling, and often are extremely unpleasant. A few of the known groups of fungus diseases will be described, in the order of, first, those that are superficial, affecting only the outer integument, and, second, those affecting the internal organs.

DERMATOMYCOSES

A considerable number of fungi grow upon and invade the skin, hair, and nails of man and other animals, and are therefore known as dermatophytes. These seldom kill or even seriously disable the host, but may cause disfigurement, irritation, and constant, long-continued annoyance. Two typical types, of the many known, will be summarized.

Ringworm of the scalp. Children are susceptible to ringworm of the scalp caused by two species of the fungus genus Microsporum. One of these species is primarily a parasite of the hide of dogs and cats, from which it may be acquired by children; the other has man as its principal host. These fungi grow on the hairs and skin of the scalp, causing the hair to fall out, leaving scrofulous bald spots. Mycelium and spores persist for some time in the hairs and skin scales that are shed from the infected areas, and the disease is spread in this way.

This fungus ringworm probably was much more common in the days when soap was a rarity than it is now. However, during the 1940's, epidemics of ringworm of the scalp were prevalent among school children in many communities of the United States, and these constituted a real public health problem. One of the complicating factors is that the fungi responsible for the ringworm may grow on and in the hair and skin of the scalp of some individuals without producing any visible infection, and can be detected only by special methods. Established infections usually require treatment for a period of months or years before they can be entirely cleared up, but they often disappear of themselves when the age of puberty is reached. Adults seldom are infected by these fungi.

Athlete's foot. Fungus infections of those portions of the skin that regularly are moist are very widespread, and some of them are extremely persistent, an unpleasant character that so many of the fungus diseases of man share. One of the most prevalent infections of this type is athlete's foot, caused by several different species of Trichophyton. These are likely to abound where groups of people use showers or similar facilities in common — i.e., where they walk barefoot over floors contaminated by the fungus from infected feet. Once established, such infections may persist for years, and at times they are complicated by bacterial infections that accompany or follow them.

The fungi which cause these infections can be grown readily on various agar media, and it is possible that they occur as saprophytes in the soil. This assumption is supported by the fact that certain species of them are almost restricted to certain geographic regions, or at least are much more prevalent in certain regions than in others. Possibly the fungi grow either saprophytically or parasitically upon various microscopic members of the soil fauna, and man may be only an incidental host. We do not yet know all we need to know, or should like to know, about the biology and mode of life of these common fungi. Our knowledge of them has increased greatly in the last few decades, but there still is room for productive research in the field.

ACTINOMYCOSES

The actinomycetes are a common, but in some ways peculiar, group of organisms. Some actinomycetes closely resemble certain bacteria, especially Mycobacterium, the tubercle bacillus, which causes tuberculosis. Yet under certain conditions these actinomycetes produce short strands of mycelium, like a fungus. This mycelium, or these mycelium-like strands, often fragment into single cells that then continue to multiply by fission, like a bacterium. These mostly bacteria-like forms, in the genus *Actinomyces*, cause several widely distributed and sometimes serious diseases of animals, from lumpy jaw of cattle to tuberculosis-like infections of man.

Many actinomycetes, on the other hand, grow generally in the form of fine mycelium, on which are produced coiled chains of small spores. These spores usually germinate to form mycelium, but may occasionally increase by fission as do the bacteria. These mostly fungus-like forms, chiefly in the two genera Nocardia and Streptomyces, are common in soil and decaying vegetation. Some species of Nocardia cause infections in man, and some species of Streptomyces infect plants.

Naturally, different investigators coming in contact with different portions of this large and varied group have had different opinions about the group as a whole and about the individual genera and species within it. Only recently have the medical bacteriologists, mycologists, and soil microbiologists begun to bring some clarity and order into the actinomycetes.

Lumpy jaw and related infections. Lumpy jaw is a disease of wild and domestic animals by no means new, since lesions typical of present-day infections have been found in the bones of a prehistoric rhinoceros. It occurs in deer, moose, horses, cows, some other wild and domestic animals, and man. At times it causes considerable loss in cattle.

The fungus, *Actinomyces bovis*, probably occurs naturally as a saprophyte or mild parasite in the mouth of various animals, where it can be found on the gums or between the gums and teeth. As a result of wounds or of decaying teeth, the fungus may invade various portions of the surrounding tissue, most commonly the jawbone. The bone tissues are partly destroyed, and overgrowths result that produce the characteristic lumps recognized as lumpy jaw in cattle.

Man is subject to infection by the same fungus. These infections probably arise much like those in cattle: the fungus normally present in the mouth can, through wounds, abscessed teeth, and so on, invade living tissue. In man it usually invades portions of the face and neck, but may also infect the lungs, intestines, and various other internal organs. Infections in the lungs frequently result in symptoms resembling those of pneumonia and also tuberculosis. Positive diagnosis depends upon culturing the fungus from the diseased tissue or upon detecting it by microscopic examination of the tissue or of pus or other discharges from it.

Several difficulties have attended the studies of these diseases and of the fungus that causes them. *Actinomyces bovis* is not likely to be found alone in the tissues it invades; various other microorganisms — bacteria, yeasts, fungi — precede, accompany, and follow it. *Actinomyces bovis* is an almost obligate anaerobe: it can tolerate only a very low concentration of oxygen. It grows slowly in culture, and is fastidious about the ingredients of the culture medium on which it grows. This makes it difficult, when a variety of other microorganisms are present, to obtain pure cultures from tissue invaded by it. When *Actinomyces bovis* is got out in pure culture and inoculated into laboratory animals, it seldom produces typical symptoms. Repeated inoculations over a period of time seem necessary before infection results. The animal apparently must first become sensitized to the fungus. Moreover *Actinomyces bovis* occurs rather frequently in the mouths of seemingly healthy individuals, so that its mere presence is not evidence of disease. Small wonder that some of the mycoses caused by it have been rather difficult to work out.

Madura foot. This disease, so named from the region in India where cases of it first were found, is a fungus infection of tissues of the foot and lower leg. Overgrowths result that may cause slow, progressive disablement, and sometimes death. Several different fungi may be involved in this, but species of the genus Nocardia, an actinomycete closely resembling *Actinomyces bovis*, are likely to be a principal cause. Unlike *Actinomyces bovis*, Nocardia may occur as a saprophyte in the soil, at least in tropical regions, and those who go barefoot may become infected by this fungus through wounds on the feet. Even in those regions of the tropics where it mainly occurs, it is not a common disease; but it can be an exceedingly unpleasant one.

COCCIDIOIDOMYCOSIS

Coccidioidomycosis is the rather formidable name applied to those diseases caused by a fungus known as Coccidioides. One and the same fungus causes several different diseases, depending on the portion of the body infected.

Coccidioidomycosis was first described in Argentina, shortly after 1890. A few years later, additional cases of it were found in the San Joaquin Valley in California. Subsequently it was found to be rather generally present in the dry regions of the southwestern United States, from western Texas to central California. It is especially prevalent in the San Joaquin Valley, around Phoenix and Tucson in Arizona, and in a few other localities of that general region, but is encountered throughout the area. The nonfatal and more common form of the disease is known variously as San Joaquin fever, valley fever, and desert rheumatism. The symptoms that accompany it include various aches and pains, fever, chills, and cough — the general pattern being similar to the symptoms of influenza and pneumonia. These symptoms arise from invasion of the lung tissues by the fungus *Coccidioides immitis*. This infection of the lungs may, after a time, clear up, and the person regain normal health. Occasionally, however, the fungus spreads from the infected lungs throughout the body, causing lesions and abscesses in the bones, joints, and various internal organs. This form of the disease is highly fatal.

The disease has a number of interesting aspects. At one time it was thought to be very rare — from 1893 to 1931 only 254 cases were reported in California, most of them fatal, and almost none were found elsewhere, outside of the two original ones in Argentina. In the late thirties it was found that most of the residents tested in the area where the disease was endemic reacted positively when an antigen prepared from the fungus was injected under their skin. Few people in other areas reacted positively to such a test. This was fairly conclusive evidence that a large majority of the people who lived in the area where the disease is known, had been infected by the fungus at some time or other. In other words, most of the inhabitants of the rich San Joaquin Valley and some of the neighboring areas are infected with the fungus. In most of them the symptoms of infection are so mild or so transitory that they go unnoticed. In only an occasional person does the infection produce symptoms severe enough to require medical attention. In most of these the infection remains localized in the lungs, and in only a small proportion does it result in serious illness or death. These facts and findings, acquired by rather intensive investigations over a period of some years, helped to remove some of the mystery from the disease. Sufficient mystery nevertheless remained.

One of the puzzles was the peculiar distribution of the disease. Why should it be limited almost exclusively to the arid Southwest? Infected people who leave that region and travel about do not set up new centers of infection. It does not spread from man to man. People infected with it are found here and there in the country, to be sure, but most of them have lived for some time in the region where the disease is endemic, or they have become infected from materials originating there. The disease does not spread readily, if at all, from one person to another; so the riddle of why the disease should be limited to the Southwest was linked with the question of how infection occurred: Where did the inoculum, the fungus spores, come from to infect the lungs of man? One obvious possibility was that, in this arid region with its seasonal rainfall and frequent dust storms, the fungus was inhaled along with the dust from the air. Attempts were made to isolate the fungus from the soil. It was got from soil occasionally, but never in sufficient quantity or with sufficient generality to make it seem that the fungus was growing regularly in the soil, as a regular member of the soil microflora. It was, however, isolated rather regularly from the soil in and near the burrows of certain rodents. With this lead, a couple of well-trained and acute investigators found that certain species of rodents common to the Southwest were rather heavily infected with this fungus. Fifteen per cent of three species of pocket mice and seventeen per cent of one species of kangaroo rat, both of which live in burrows in the soil there, had infections of the lungs caused by this fungus, infections that were similar in many respects to those encountered in human beings.

These rodents appear to be sufficiently tolerant of the disease, or sufficiently resistant to it, so that infected individuals are not killed quickly by it. They can contaminate the soil in and near their burrows for a considerable time with the fungus. In the dry season this contaminated soil is picked up by the wind, inhaled by people, and those susceptible become infected. Most of them become infected at some time or other, as stated above, but in only a few cases does infection result in obvious or serious disease. It now seems likely that these rodents are the primary reservoir of inoculum, possibly the only one. Man probably is only an incidental host.

When the army established a number of air fields in the San Joaquin Valley in 1941, the medical authorities did not have all of the above facts available, because some of these facts were not then known. They did have enough information available to give them good clues on which they could base control measures. They had roads and runways paved where this was possible, established grassy turf where possible, and treated playing fields and paths with oil. All of these public health measures were aimed at reducing the amount of dust in the air. To be sure, all these measures combined could only reduce local dust contamination, and could have no effect on the dust from outside the immediate area. They served to reduce by from one half to two thirds the infections caused by this fungus. Apparently the fungus does not survive far travel by air, since if it did we could expect infections to be common wherever the dust from the Southwest is carried by the wind.

THRUSH AND RELATED DISEASES

Candida is a yeastlike fungus, with usually sparse mycelium, on which masses of budding cells are produced. It is half filamentous fungus, half yeast. Some species of Candida are common and abundant on the outside of seeds such as corn and barley, some occur in soil, others on fruits. One of these, *Candida albicans,* causes various diseases in man. Of these diseases some are superficial infections of the skin, usually between the fingers and toes, much like the infections caused by Trichophyton and its relatives. Some are superficial infections of the mucous membranes of the mouth and throat. Some are systemic or generalized infections of the lungs, liver, lymph nodes, spleen, and intestines, which occasionally are fatal.

Thrush is an infection of the mucous membranes of the mouth and throat by *Candida albicans.* It is primarily a disease of infants and younger children, and apparently is much less prevalent now than it was in the past, when epidemics of it were common in foundlings' homes. Since *Candida albicans* is another of the fungi that normally inhabit the mouth and throat, cases of parasitism by it sometimes are difficult to diagnose. The mere presence of the fungus in an infected mouth or throat is not evidence that it is the cause of that particular infection.

Candida albicans also causes infections of the lungs, and certain of these infections closely resemble certain types of tuberculosis. These cases may be difficult to diagnose, since both the fungus and the bacterium may be present in both seemingly healthy and obviously diseased individuals. The situation is further complicated by the fact that different strains of the fungus exist, some pathogenic and some sacrophytic, and by the fact that individuals who in good health may be resistant to infection by those strains of *Candida albicans* normally present in mouth, throat, and intestines, may, as a result of weakening illness, become subject to attack by them.

The account in the preceding two chapters is only a summary of a few of the many diseases of man and other animals caused by fungi. It includes examples from a number of the common and important types of fungus infections of various members of the animal kingdom, but does not cover the whole range by any means. It should serve to show that fungi are important in the lives of lower forms of life, from nematodes to insects, and that they are of some significance in the well-being and survival of various higher animals, including man. Some of these fungus diseases of man have their complex aspects, as do some of the fungus diseases of plants. The principles we establish by basic studies of how fungi operate, how they grow, survive, and reproduce, even if these studies are not aimed at any immediately practical end, are often essential to our understanding of how these fungi behave in any given instance. In other words, basic research on fungi, and on all aspects of fungi, is essential to our understanding of them. We must live with them, including those that infect and parasitize our bodies, as we must live with so many other plants and animals on the face of the earth. The more we learn about them, the better chance we have to reduce or control the evils they cause, and to promote any beneficial aspects they may have.

Additional Reading

G. C. Andrews, *Diseases of the Skin*, 3d ed. Philadelphia: W. B. Saunders Company, 1946. 937 pp.

R. J. Dubos, ed., *Bacterial and Mycotic Infections of Man*. Philadelphia: J. B. Lippincott Company, 1948. 785 pp.

E. C. Large, *The Advance of the Fungi*. New York: Henry Holt & Company, 1940. 488 pp.

C. E. Skinner, C. W. Emmons, and H. M. Tsuchiya, *Henrice's Molds, Yeasts and Actinomycetes*, 3d ed. New York: John Wiley & Sons, 1947. 409 pp.

CHAPTER 10

Fungi Exploited Industrially

IN THE last few decades some fungi have become big business in the Western Hemisphere, but that products useful to man could be got from molds is no new discovery. The surprising thing is not that a few fungi have been put to work for us but, that out of the tens of thousands of common fungi around us, we have had the wit to exploit only a scant dozen for public good and private profit. All of these have been hit upon purely by accident.

The Orientals were far ahead of us in this, as they were in so many other things. Soy sauce, saki, and cheeselike foods made from soybeans and rice were being manufactured with the aid of molds in the Orient more than two thousand years ago. Today, one of the large regional laboratories of the United States Department of Agriculture has its inquiring research fingers deep in soy sauce, with the idea of converting this ancient Oriental mystery into a modern mycological process. Out of such research may come more important things than just a recipe for making soy sauce. They may find out how fungi can convert soybeans or other agricultural products into new feeds and foods and other useful things.

We may now be far from an Oriental economy (essentially one in which little protein food is got from animals, much from plants), but we are getting closer to it. If you doubt that, look up the statistics. In the United States in the last sixty years soybean production has increased tremendously. In 1900 we ate practically no soybeans in any form, but now from 10 to 15 per cent of the large soybean crop is going into human food — everything from sausage and canned soup to breakfast cereals. Given the solution of a couple of tough technological problems, the use of soybeans in food will increase even more rapidly. During the last twenty years the population of the United States has increased at an ever accelerating rate, in 1960 amounting to 180 million people. If we have another fifty years of such population growth, we shall almost inevitably have to get much of our protein from plants, soybeans among them. In the Orient fungi have long helped to convert soybeans and rice into a variety of palatable, storable, nutritious human foods. And so the present research on soy sauce may affect our future habits, health, and even our prejudices in ways now unforeseen.

We Occidentals have also discovered a number of uses for some of the common fungi. Some cheeses are ripened with the aid of fungi. Citric and gluconic acids, of considerable commercial importance, as well as some organic acids of lesser importance, are produced chiefly or solely with the aid of fungi. A process recently has been developed for the successful production from a fungus of diastase, or amylase, an enzyme which converts starch into sugars that the yeasts can ferment into alcohol. Fungi furnish us with drugs such as ergotine, penicillin, streptomycin, aureomycin, terramycin. The waste fungus material from some of these fermentations is rich in growth-promoting factors that serve as a valuable adjunct to animal feeds.

That there is such a field as "industrial mycology" will be news to many. That courses in the subject are given in various institutions, and textbooks written on it, constitutes evidence that molds are beginning to be recognized as a factor of some importance in our industrial life. Fungi are certain to become more important in the future, as research opens up new possibilities of exploiting them. The following account does not cover the whole field, but aims to introduce the reader to some of the principal ways in which molds other than yeasts have become big business.

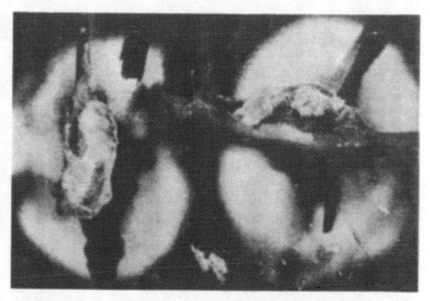

Plate 12. Two adult flies that have been invaded and killed by a fungus related to Empusa. The dying flies settled on twigs, which were transferred to a glass slide and kept in a moist chamber overnight. The white circles on the slide were formed by spores shot off by the fungus.

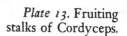

Plate 13. Fruiting stalks of Cordyceps.

Plate 14. Some of the ingredients and equipment used in preparing culture media. In the back row, common salt, malt extract, a prescription bottle containing prepared agar medium, and a can of powdered agar. In the front row, a Bunsen burner in front of a petri dish in which wheat seeds have been cultured on malt-salt agar, with fungi growing from the seeds, a stack of petri dishes ready to have melted agar poured into them, and a petri dish containing dry powdered agar. With no more equipment than this, plus a pressure cooker, anyone can culture a great variety of fungi.

The yeasts are fungi too, but they are almost a special field in themselves and will be touched on only lightly.

Foods from Fungi

Cheeses. Not all moldy food is garbage. It is true that moldy bread or meat or fruit is undesirable, and enough mold in milk or butter is considered evidence of filth. Yet some of the most delectable cheeses are ripened and given their characteristic flavors and textures by some of the same fungi that spoil other foods. Of these cheeses, the Roquefort type is perhaps the best known.

Roquefort was so named from the region in France where this general type of cheese first was made, several hundred years ago. At that time the sheep-milk curd, the raw material of the cheese, was not deliberately inoculated with mold. Between the udder of the sheep, goat, or cow and the cheese-ripening rooms, milk and cheese curds are likely to pick up a random sample of the materials present in the barns, dairies, and cheese-making rooms. The book *American Chamber of Horrors* lists some of the strange things occasionally encountered in milk and cream in some of the dairy regions of the United States in the not too distant past. There is no reason to suppose that the farmers around Roquefort, France, a couple of hundred years ago, were noted for undue cleanliness. It is likely that the sheep's milk which arrived at the factories contained not only the usual high quota of molds, bacteria, and other microflora, but also assorted debris and offal. The larger objects were strained out, of course, but the bacteria and molds remained. The milk was not pasteurized, and so the newly made cheese was literally crawling with microorganisms, all fighting for survival.

The conditions under which the cheese was stored determined which of the many kinds of fungi and bacteria present would eventually dominate and take over, and this largely determined the character and quality of the final cheese. In the Roquefort region the newly made sheep-milk cheese was stored in limestone caves. The percolation of water through the rocks of the caves kept the air cool and humid. Under these conditions one mold, since named *Penicillium roquefortii* (a close relative of *Penicillium notatum*, from which penicillin is made) often became dominant in the cheese. After several months of storage, this fungus growing in the cheese gave it a soft texture and wonderful tangy flavor. That is, the mold grew in and partly digested the milk curd and fat. Or rotted it. To the layman it sounds better to say that the cheese was ripened by the mold, rather than rotted, but the difference is only one of words. The fungus as it grew in and ripened, or rotted, the cheese, produced masses of bluish-green spores distributed in irregular veins and pockets through the curd, and these spore masses gave the cheese its characteristic mottled appearance.

The first production of this cheese was just a happy accident. The cheese makers were just trying to make the same cheese their forefathers had made, full of filth and flavor and nourishment. They happened to store the curd where it almost inevitably ripened into the characteristic product later known as Roquefort cheese. They knew nothing about fungi, and doubtless would have considered any study of molds in cheese a deplorable waste of time and money. The making of this cheese became an art, and soon became surrounded with the usual hocus pocus of such arts. Those lucky enough to have the right conditions – including a heavy enough contamination by the right fungus – for ripening the cheese were able to make good Roquefort some of the time. They lost a portion of the cheeses through other molds or certain kinds of bacteria getting in and taking over and inducing the wrong kind of spoilage. The cheeses they produced varied greatly in texture, flavor, and palatability, and probably also in the amount and kind of extraneous filth they contained. The quality depended partly on what was in the milk from which they started, and what got into it later. They could never be too sure, when they made a hundred cheeses and put them away to ripen for a few months, just how many were going to turn out good. In other words, they did not know how to make Roquefort cheese.

Sometimes they made it and sometimes they didn't. When they did, it was an art.

Modern research has converted the manufacture of this type of cheese into a science. If this has removed some of the romance from the process, it has also removed much of the dirt from the cheese, and removed much of the risk of failure. Shortly after 1900 Charles Thom, mycologist with the United States Department of Agriculture, began to study molds in cheese. The work was undertaken primarily because some of the cheese makers in the eastern part of the United States suffered some loss from undesirable molds growing in and on ordinary cheddar cheeses. These fungi spotted the cheeses and gave them off-flavors, and Thom set out to find out what he could about the molds on and in cheese.

Before long he and his co-workers found out that one of these fungi was responsible for the characteristic flavor, color, and texture of Roquefort cheese. Up to then this cheese had been a product peculiar to Roquefort, France, although similar types, ripened by the same fungus, were made in Italy, Belgium, England, and elsewhere. Once the fungus was known, and with the general know-how in cheese making that prevailed in various parts of the United States, plus a goodly amount of research, the making of Roquefort-type or blue cheeses became an industry of respectable proportions on this side of the Atlantic. Cheeses of this type now made in Iowa, Minnesota, Washington, Wisconsin, and a few other states equal or excel the imported product in flavor and texture. If you question that, get a number of samples of cheeses of the Roquefort type from various places in the United States, and a couple from France. Set them out, unlabeled, and let your connoisseur friends decide. For a fair test, samples should be replicated several times, and each taster should rate each one, to avoid the errors inherent in chance sampling.

Good Roquefort-type cheeses were not made without some rather basic and thorough research. Many factors had to be studied critically. The particular strain of fungus used, the quantity of spores applied, how they are applied, the amount of salt added and when, the number and size of holes punched in the cheese to give the mold air, but not too much air, the temperature and humidity and time of storage — all these and many more affect the quality of the final product. As one result of the extensive and intensive research done mostly in a few agricultural experiment stations, the making of this cheese is a controlled process. The blue cheese of different makers often differs, but by choice, not happenstance. We can now buy blue cheese hard and crumbly, soft and smeary, having different degrees of sharpness or bitiness to suit our individual tastes. Where once there was available only the imported product, so expensive that only the moderately well-off could afford it, we now get these good cheeses at a reasonable price, and with a high assurance that they are free from the hairs and excrement of rodents and ruminants, and free of the germs of tuberculosis and undulant fever. For those who relish not only good food but clean food as well, the conversion of this process from a hit-or-miss art into a science or technology should be welcome. The dairy industry has not complained.

Camembert is another fungus-ripened cheese. Like Roquefort, it was named from the region in France where it first was produced. Thom, in his studies of fungi in cheese shortly after 1900, found the following molds in Camembert cheeses bought in the open market: six species of Penicillium; two or three species of Aspergillus; *Oidium lactis* (an organism almost universally present in milk); *Cladosporium herbarum*; *Monilia candida*; and Mucor, Fusarium, Cephalosporium, and Acrostalagmus. While these technical names of the fungi may mean little to you, they should indicate that a variety of different kinds of fungi were present in the cheese tested. Of these several molds, two were abundant on or in every sample. These two were a species of Penicillium, since named *Penicillium camemberti*, and *Oidium lactis*.

Thom was able to produce the texture typical of Camembert cheese by inoculating the raw curd with *Penicillium camemberti* alone, but such cheeses were flat and flavorless. Cheeses inoculated first with *Penicillium camemberti* and allowed to half-ripen, then inoculated with *Oidium lactis*, developed the typical Cam-

embert flavor within a week. The few kinds of bacteria sometimes present on or in such cheeses apparently had no more than
an incidental effect on the final product. It was found advanta-
geous to inoculate these cheeses, a day after they were made, with
spores of *Penicillium camemberti*. *Oidium lactis* was so universally present in the cheese-making establishments that no inoculation was necessary, although different strains of this fungus
were able to produce different and characteristic flavors, and thus
might be useful in making different types of this savory cheese.

Relatively little research work has been done with Camembert
and some of the other cheeses ripened with fungi. There is room
here for much more research. Will other kinds of molds, or other
strains of those now used, ripen milk curd into still other kinds
of cheese, of new and more attractive flavors? Almost certainly
they will, but no one has yet seen fit to explore the possibilities.
What of the possibility of inoculating soybean curd, or pressed
cakes of certain cereal grains, or bran, or cottonseed meal, with
various fungi, singly and in combination, just to see what sort of
new foods we could produce? Work of this sort in all probability
would lead to more profitable use of some of our agricultural
products, and in the long run might turn out to be of major
significance in our economy. Here again a research program
should be established not just to find immediate answers to immediate problems, but to prepare for the future. Fifty years from
now a backlog of research of this nature might well mean the
difference between an adequate diet and partial starvation for a
portion of our population, especially with a crop failure in the
Midwest breadbasket. It might also lead to a more abundant life
even in an economy of plenty, should such an economy prevail
anywhere in the world fifty or a hundred years from now.

Cultivated mushrooms. We do not ordinarily think of the
mushrooms we buy in a store as molds, yet there is little essential
difference between common molds and common mushrooms.
Both molds and mushrooms are fungi. Mushrooms just happen to
be big enough to be seen, and good enough to be attractive as
food. In the Western Hemisphere mushrooms have been grown

for about four hundred years. Of the thousands of different kinds of wild mushrooms, only a few hundred are big enough, of good enough flavor or keeping quality, or common enough to be used much for food. Of these, few can be cultivated. Of these few, only one is grown commonly in Europe and America. *Agaricus campestris* by name, it grows in nature in grassy fields or on compost heaps. Actually it is even a little bit more complicated than that, since not any *Agaricus campestris* you find can be cultivated. The common *Agaricus campestris* on lawns and grassy places has four spores per basidium. A variety of it found almost only on compost heaps or manure piles has only two spores per basidium — some of the Europeans call it *Agaricus bisporiger* — and only this two-spored one can be cultivated.

This mushroom was first grown in France, in limestone caves in and near Paris, caves made by quarrying stone to build the city. It can be grown successfully and on a large scale only where the temperature is uniform and not too high, the relative humidity of the air uniform and not too low, and there is a constant but gentle circulation of air. The caves in and near Paris met these requirements excellently.

Until 1900 there were almost no mushrooms cultivated in the United States, although some were grown in England. France had a virtual monopoly on mushroom production and, naturally, exploited it. It was said to be too difficult and subtle an art to be mastered by anyone but a Frenchman, and even Frenchmen sometimes ran into difficulties. Here again, men in the United States Department of Agriculture helped provide a background of research that turned this difficult art into a science and made the present large mushroom-growing industry possible. Growing mushrooms is still not very easy. If commercial mushroom production tends to be monopolized by a few hundred growers, the blame can be placed chiefly on the mushroom itself. Anyone who is willing to follow available directions can grow mushrooms. Few can grow them in sufficient quantity and regularly enough to make money at it.

The process is about as follows: The raw material is horse manure (a compost mainly of corn fodder has recently been developed at Pennsylvania State College, but is not yet in general use). It makes some difference whether the manure comes from a cavalry horse, artillery horse, farm horse, or purely decorative horse, or race horse. This may sound ridiculous, but it happens to be so. The amount of straw, shavings, or other material in the manure also makes some difference. It is essential that the manure be fresh.

The fresh manure is piled on the ground, and the size and shape of the pile affect the results. When fresh manure is properly piled it will heat, because of the growth of molds and bacteria in it. It should attain a temperature of about 140 degrees Fahrenheit, and remain at this temperature for about ten days. Sometimes it doesn't get hot enough, sometimes it gets too hot, sometimes it is just right. After this preliminary heating, the manure is repiled and allowed to heat again. This heating not only conditions the manure and makes it favorable for the growth of *Agaricus campestris*, but also kills off many insects, injurious or competing fungi, bacteria, nematodes, and other flora and fauna that might interfere with the mushrooms. Proper heating of the manure is absolutely essential to successful mushroom growing.

Once properly cured, the manure has been transformed into a rather pleasant-smelling compost of crumbly texture. It is then piled in beds. These may be only six inches deep, several feet wide, and as long as the cave or house or quantity of manure allows, or may be only eighteen inches wide and two feet high. A more recent development is to place the compost in shallow wooden trays equipped with short legs so that they may be stacked on top of one another. Once in beds or trays, the compost is inoculated with the mycelium or spawn of the mushroom. This can be grown easily by anyone who knows how to culture fungi in the laboratory, but usually it is obtained from commercial spawn producers who grow it by a simple but patented process. A process, by the way, that, long before it was patented, was generally used in plant pathology laboratories to grow a variety of fungi.

The spawn is scattered on the compost and, after a couple of weeks, mycelium of the fungus permeates the bed. Then the bed is cased, or covered with black soil. If the manure was of the right sort and properly composted, if the spawn was good and the casing soil right, and if conditions have been favorable, mushrooms begin to appear after from five to six weeks. Successive flushes, or crops, appear every week or so for from six weeks to three months, by which time the compost is exhausted. It is then removed, replaced, and the process repeated.

The growing of mushrooms is attended with many hazards. A grower may compost a few dozen cubic yards of manure, turn it, place it in beds, inoculate it with spawn, case it with soil, wait several weeks, and no mushrooms appear. So the beds must be discarded. Also mushrooms, like other plants, are subject to a variety of diseases and insect pests, and these sometimes ruin promising crops. A sharp fluctuation in temperature or humidity during the small hours of the night may result in partial crop failure. Variations in the wholesale price of mushrooms often are sharp and unpredictable.

These factors are mentioned only because many otherwise apparently reasonable citizens have got the idea that they can go into the business of growing mushrooms in a shed or basement and make a big, quick, and easy profit. It isn't so simple as that. Mushrooms can be cultivated on a small scale in the basement if one cares to devote time and energy to that type of agriculture. Detailed directions for growing them can be got from commercial growers and from the United States Department of Agriculture. But one starting out with a few bushels of manure in his basement is not likely to compete with the commercial growers who have been in the business for decades, who produce tons of mushrooms daily, and who even have their own research staff to cope with the new problems that constantly arise.

While on the subject of cultivated mushrooms, morels should be mentioned. These spring mushrooms are among the elite of the edible fungi — crisp, savory, and enticing, one of nature's superior delicacies. In a recently published mycology text it is stated that morels can be grown much as cultivated mushrooms are grown. Would that this were so. Actually there is no authentic record of a morel's ever having been artificially grown.

True, there is an account, in a century-old issue of the *Gardeners' Chronicle*, in which a Frenchman claimed to have grown morels at will in flower pots. Having happened upon this account in desultory reading in my scientific youth, I became somewhat excited about it. It is well known that the spores of morels will germinate, and the mycelium grow readily and rapidly, on various media used to culture fungi in the laboratory. At the time I chanced on this old account, we happened to have a number of cultures of morels in the laboratory. Following the directions given so explicitly by this ancient Gaul, we inoculated mixtures of soil, bark, and leaf mold with morel mycelium. The mycelium grew wonderfully, but not a single morel was produced. Some of the pots full of mycelium we planted out in our arboretum, under various kinds of trees. In the ten years that have elapsed, not a single morel has appeared, although they come up every spring in an orchard a few yards away. Later we found out that we were not the only ones who had tried to grow morels; much more expert and knowing mushroom cultivators had also tried it, and failed.

The reason for our and other attempts to grow morels is that they command about twice the price of ordinary cultivated mushrooms. They are worth it. Morels appear in profusion in a great diversity of environments throughout the whole temperate region of the northern hemisphere during early spring. This is evidence that the conditions necessary for the production of morels in nature prevail rather widely. Someone may eventually discover how to cultivate them. Probably an amateur botanist or mycologist or gardener who does not know enough to realize that a certain approach is logically impossible. The one who does cultivate them will have a gold mine, and one that can be exploited for a long time. There is money in morels. But first, grow them.

Enzymes and Acids from Fungi

Fungus diastase. European and American countries have traditionally used barley malt to convert the starch of cereal grains, potatoes, or what have you into sugars that yeasts can ferment into alcohol. In the Orient a fungus has been used for this purpose since about 1800 B.C. The ancient Japanese sake, or saki, is, like the present sake, a rice wine made with the aid of a fungus. Rice is washed, moistened, piled up, and inoculated with a fungus. Within a few days the fungus has converted most of the starch in the rice into fermentable sugars. The rice is then put in vats, yeast is added, and the resulting fermentation produces saki. The fungus involved is *Aspergillus oryzae*, or *Aspergillus flavus*. It is present on most agricultural seeds, and often is responsible for the spoilage of seeds stored in bulk, as we have already seen in Chapter 7.

Given just the right water content in the seeds, *Aspergillus oryzae* will take over, to the practical exclusion of everything else. Long before 1800 B.C. the Japanese had found by experience the proper conditions to allow this fungus to grow essentially as a pure culture on the rice being prepared for fermentation. For a long time the manufacture of this wine was a household art, but before the Western world knew its way around with fermentation and wine production, the Japanese had developed the fermentation of sake to a fairly exact science. About two hundred years before the time of Pasteur, they even heated the finished wine to prevent later spoilage. That is, they pasteurized their wine a couple of centuries before Pasteur hit upon the heating process to preserve French wines from subsequent spoilage, and so gave his name to the heating process now generally used to eliminate some undesirable microorganisms from milk and many other food products.

In the early 1900's Takamine, a Japanese scientist with some promotional ability, spent a few years in the United States trying to develop the process of converting the starch of cereal seeds into fermentable sugars by the use of the fungus traditionally used in the preparation of rice for sake. Several distilleries in the United States and Canada tried his process. It apparently resulted in a product with a strong moldy flavor, and so never gained much favor.

In recent years interest in the process has been revived. Alcohol is used in many products other than the demon rum, and in some of these a slight flavor of mold is not objectionable. Work with different strains of the fungus, with different methods of growing it, and with other steps in the process have served to make the production of fungus diastase a rather large and rapidly growing industry. *Aspergillus oryzae* is grown on bran, a relatively cheap by-product of the wheat-milling industry, and on this substrate it produces large quantities of diastase. Diastase that is just a little bit more efficient than that of barley malt in converting starchy grains into fermentable sugars. Malting barley is an expensive, variable, and often nonstandardizable product. Its quality varies with the variety of barley, the amount of fungus infection on the seed, the soil in which the barley was grown, the weather, and various other factors. The fungus-produced diastase is a far more controllable and standardizable product. It probably never will replace barley malt for certain purposes, but in the last few years it has made surprising headway, and is now being produced commercially in quantities equal to several hundred thousand bushels of malting barley per year. Alcohol is a source of energy of some present and more potential value. We may well be heating our homes and running our machines on alcohol long before we heat and run them with atomic energy. If we do, a fungus will almost certainly be an essential part of converting the plant starch into sugar, as well as the sugar into alcohol.

Citric acid. Citric acid may not be a vital part of our civilization, but we do use a few thousand tons of it every year. About 65 per cent of it goes into drugs and medicines of various kinds, somewhat more than 10 per cent is used in foods and soft drinks, and the rest in silver plating, engraving, and the dyeing and printing of cloth. Other uses for it may be found as it becomes still cheaper.

Citric acid was first isolated in pure form in 1784, long before any commercial use was known for it. Up to 1922 Italy produced about 90 per cent of the world's supply; it was got principally from low-grade lemons. The major part of this production was exported to the United States, in the form of calcium citrate. For this salt we paid about 53 cents a pound in 1914; but by 1919 it cost us a bit more than a dollar a pound, and a year's supply cost us between two and three million dollars.

In 1893 citric acid was isolated from a mold, and it was no secret that several molds were able to produce appreciable quantities of it. If a commercial process for such production could be developed, it might pay off handsomely. Work on the problem was undertaken, and within a few years a successful process was developed — chiefly by one New York manufacturer of pharmaceuticals. In 1925 the price was down to 46 cents a pound, in 1935 it had dropped to 28 cents, and European countries organized a cartel which practically eliminated this low-priced, mold-produced citric acid from their market. By 1944 we were producing around 13,000 tons of calcium citrate per year, more than 90 per cent of it from a mold.

The mold used is a selected strain of *Aspergillus niger*, a common black fungus that is prevalent on decaying vegetation, and is a common weed in laboratory cultures. Not every strain of this mold will produce citric acid, and few will produce it in commercial quantities. In practice, spores of the fungus are sown on the surface of a liquid medium in shallow metal or enameled pans. The liquid is sufficiently acid to keep out most competing organisms, and it is not necessary to maintain complete sterility, as in the production of penicillin. Within a few days the mold forms a thick mat of mycelium on the surface of the liquid, and by then has excreted as much citric acid into the liquid as it is able to. The liquid is then drained off, the acid precipitated from it as calcium citrate, and the salt is purified, packaged, and sold.

Details of the process are mostly the secret of the one drug company that has developed the process, although the most essential secret probably is the know-how that comes from years of intimate work with a balky and unpredictable mold. It is estimated that literally acres of pans were necessary to produce the amount of calcium citrate being produced by this company in the early forties. This same concern, by the way, was among the first anywhere in the world to produce penicillin on a large scale, no doubt partly as a result of its work with this other and similar mold fermentation. For several years it produced more than half the penicillin manufactured in the world, and it still is responsible for a large share of penicillin. Such a monopoly is not the result of vicious capitalism, but of technological skill and information.

Attempts have been made to produce citric acid by submerged fermentation, in large vats. On a laboratory scale this has been done at the Northern Regional Research Laboratory at Peoria, Illinois, and at other places. Almost certainly it will be done soon on a commercial scale. For all we know, citric acid may be produced in that way commercially now. If so, it means that citric acid will become still cheaper and more plentiful and will find its way into new products, such as plastics, from which it previously was barred by its relatively high cost.

Gluconic acid. Gluconic acid was first found as a product of mold fermentation in 1922, more or less as an accident. An alert research worker recognized that not all the acid produced by the citric acid mold was citric acid, and one of the other acids he isolated was gluconic acid. The compound had been known before it was isolated from molds, and various industrial uses had been suggested for it. For one thing, calcium gluconate is an excellent source of calcium for pregnant mothers and nursing children — superior to most other forms of calcium. It also has produced excellent cures when injected into cows suffering from milk fever. It can be manufactured chemically from dextrose, though the process is rather expensive. Since the discovery that it is produced by the same fungus that produced citric acid, a number of other kinds of fungi have been found to produce it. High-producing strains of some of these have been selected, and a submerged fermentation process has been developed that increases the yield. In 1945 some 871,000 pounds of calcium gluconate were produced in the United States, most of it from fungus fermentation. While not now a major industrial product, it is a valuable specialty, and doubtless as it becomes more plentiful and cheaper, we shall find more uses for it.

Drugs from Fungi

Ergot and ergotine. Although various fungi were used in ancient medicine, ergot was one of the first fungus drugs of value. A German book published in 1582 mentions ergot as a useful and proven aid to accelerate labor in childbirth. During the eighteenth century crude ergot was commonly administered by midwives in various European countries to their patients, or victims, as an aid in childbirth. They merely picked ergot kernels from harvested rye, ground them into a powder, and administered the powder, usually in special decoctions to add flavor, color, a medicinal odor, and an alcoholic kick. In the 1800's ergot became rather generally accepted by the medical profession as a drug of some value.

The ergot of commerce comes from a fungus that infects the flowers of rye and related grasses, and replaces the seed with a long, black spur of fungus tissue. It is not uncommon in wild rice; the writer has picked more than a teaspoonful from a pound of wild rice bought at a trading post. That from rye is supposed to contain more of the physiologically active ergotine than that from other grasses; even in rye, however, the quantity and quality of the essential drug will vary with time and geography and other factors. At one time most of our ergot was imported from Spain, but in the last decade or more much of it has come from domestic sources. Some research has been done on the production of artificial epidemics of ergot in experimental plots of rye, the idea being to use the rye to grow a crop of ergot. Some of these methods have appeared to be successful on a small scale, but, so far as is known, most or all of our ergot still comes from naturally infected rye.

For centuries before ergot gained its hard-won reputation as an acceptable drug, it was a minor and sometimes a major scourge of the common people in many European countries. In the Middle Ages rye became extensively cultivated in Europe, and in many of the northern regions it was the chief cereal grown for food. Rye is much more susceptible to ergot than the other common cereals because the flowers are open-pollinated and remain exposed to infection for days. Much of northern Europe is fog-bound and drizzly during the flowering season of rye, and this means that in some years and in some regions ergot is bound to be plentiful in rye. When rye was the staple food of the people there, it was inevitable that the people should occasionally consume a good deal of ergot. Especially the lower classes. Sometimes the grain was cleaned, the nobility and clergy took the clean grain, and the ergoty stuff was left for lower-quality folks who had only done the work of raising it.

Taken consistently, in even small doses, ergot eventually produces convulsions, gangrene, and painful death. Old accounts describe the sloughing off of fingers, toes, and even of arms and legs, from eating ergoty bread. In two provinces of France in the year 994, more than forty thousand people died of ergotism. Other severe epidemics of ergotism occurred in France in 1039, 1085, and in the next century. Vitamin deficiencies were involved in the picture too, but for centuries ergotism was a fearsome thing.

There have been almost no serious outbreaks of ergotism anywhere in the world for more than a hundred years. Not because ergot no longer is present in grain, but because ergot finally was recognized as the cause of this poisoning, and methods were developed to eliminate it from the grain. That ergot no longer reaches the consumer in flour or other cereal products is due to vigilance on the part of elevator men, grain inspectors, millers, and men of the Food and Drug Administration. Many men who consider only those activities "practical" that raise a sweat or an income often do not realize how they are protected and coddled in our present society, as a result of information obtained through "impractical" research.

Penicillin. There are several excellent summaries of the penicillin story, and so the account given here will be only a résumé of what seem to be the more interesting aspects of it from a mycologist's standpoint. Almost everyone who reads newspapers now knows that penicillin was discovered because Dr. Fleming, in a London hospital, in 1929 noticed that a mold which had got into one of his culture dishes inhibited the growth of bacteria. Ten years later a group of workers at Oxford, England, isolated a substance, from the liquid in which this mold was grown, that would cure both mice and men of seemingly fatal infections by certain bacteria. They named the substance penicillin, and they suspected that if enough of it could be produced, in pure enough form, it would be of tremendous value in medicine. In 1941 two of the English workers who had pioneered in this research brought a culture of a penicillin-producing mold, plus some ideas and enthusiasm, to the United States. Work on the large-scale production, isolation, and purification of penicillin was got under way at once, first at the Northern Regional Research Laboratory at Peoria and at various commercial laboratories, and later at a number of institutions scattered throughout the United States and Canada.

To obtain penicillin, the mold is grown in a liquid nutrient medium, or beer. The penicillin is secreted into this liquid by the growing mold, and later has to be separated from all the other things in the liquid, and purified. The liquid in which the mold grows has to be free of all other organisms; that is, before it is inoculated with the mold it has to be autoclaved, or pressure-cooked, to make it sterile. One reason for this is that the air is full of bacteria, some of which are able to grow in the same liquid, and able to produce an enzyme that quickly destroys penicillin as fast as the mold produces it.

Different strains of the fungus vary greatly in the amount and kind of penicillin they produce and in the rapidity with which they produce it. A given strain of the mold will vary in productiveness according to the composition of the liquid in which it is grown, the amount of air available, the temperature, and various other factors. Also the composition of the liquid influences the ease with which the penicillin can be isolated and the purity and potency of the final product.

When penicillin was first produced commercially in this country, the mold was grown on the surface of a liquid in quart milk bottles — about half a pint of liquid per bottle. On this bottle system, a large producer had every day to empty, wash, fill, sterilize, and inoculate thousands of bottles. The strain of the mold brought over from England would grow, or at least produce penicillin, only when it was growing on the surface of a liquid; when grown submerged in the liquid in vats, it produced almost no penicillin. Men at the Peoria laboratory soon found several other strains of the mold from their own culture collection that were able to grow in large vats, and by 1944, mostly because of these new mold strains, the large-scale submerged fermentation became a reality.

All of the liquid going into a vat had to be sterile, or had to be sterilized after it was put into the vat, and it then had to be cooled before being inoculated with the mold. The mold also needs air, and so sterile air, from which all organisms had been filtered, had to be pumped through the liquid, at the rate of about a gallon of air per minute per gallon of liquid in the vat; that is, with a large vat, from 8,000 to 10,000 gallons of sterile air per minute. The liquid also had to be constantly and violently stirred if the mold was to produce a maximum amount of penicillin. There were other tough problems connected with this large-scale production that the industrial engineers had to solve, and these engineers played at least as big a part as the biologists and chemists in making penicillin available in quantity.

The mold brought over here from England was strictly a surface culture, and it yielded only a few units of penicillin per cubic centimeter of liquid, or around 2,000 to 3,000 units per quart. At a half pint of liquid per quart-sized milk bottle, this did not amount to much. Better strains of the surface culture found at the Peoria laboratory increased this to more than 100 units per cubic centimeter — an increase of from twenty to thirty times. The strain of the mold first used for submerged growth in vats produced about 50 units of penicillin per cubic centimeter of liquid, or a bit more than 50,000 per quart. As a result of research sponsored by the War Production Board in 1944, more than fifty thousand different strains of *Penicillium notatum* and related molds were tested for submerged production. One of these, which was originally picked up on a rotten cantaloupe in a Peoria grocery store by workers at the Regional Laboratory there, and from which improved strains were later selected by men at various other laboratories, was by 1945 producing around 800 to 1,000 units per cubic centimeter — an increase over the original culture of from four to five hundred times. This helped make penicillin cheap and readily available.

At the same time, research was going on in the improvement of the liquid in which the fungus was grown, and the composition of the liquid used was found to be just as important as the strain of the mold used and the fermentation conditions. Thousands of different formulas were tested, and by 1945 the recipes were being designed not only for a particular strain of fungus, but also for the particular conditions under which that strain was grown in this or that plant.

Once the fungus has produced its maximum quantity of penicillin in the liquid in which it is growing, the drug has to be extracted and purified fast. The first extraction processes got out only a fraction of the penicillin present in the liquid, and the penicillin extracted was a crude product — a yellowish brown substance of unpleasant odor, low potency, and short life. Research on these phases of the problem led to a much higher extraction of a much purer, more potent, more durable product.

In 1941 there was scarcely enough penicillin produced to treat a single patient. By 1946 the production was measured in tons and by 1951 in at least several tons per month; the potency, purity, and keeping quality of the drug were all increased many times over; and the stuff was available in many different forms, none of them expensive. This sketchy account does not even mention a number of the problems encountered in transforming penicillin from an idea into a cheap and generally available drug.

It does suggest, however, that research workers in several different fields were invaluable in contributing to the solution – workers trained in basic or applied mycology, chemistry, engineering, physiology, or other sciences.

Streptomycin. The work of producing and purifying streptomycin posed about as many problems as were encountered in the production of penicillin. Less is known about the story of streptomycin, partly because this fungus has never got out into the public domain. Dr. Waksman, of the New Jersey Agricultural Experiment Station, is the man that could tell the story, but for very understandable reasons he has had to keep it largely to himself.

The genus Streptomyces is closely related to the genus Actinomyces (some include them both in Actinomyces), and molds of these genera are common in soil, decaying vegetation, and manure. The earthy odor of soil – to some people synonymous with the simple life and wholesome virtue – is the odor of Actinomyces. You can get this odor without any toil or dirt whatever by merely keeping a culture of Actinomyces in your apartment. These organisms are half-way between the bacteria and the true molds, in a sort of taxonomic no man's land. They do not fit into a convenient pigeonhole, much to the annoyance of those who spend their time putting living things into pigeonholes, a low but necessary form of scientific endeavor. One of these molds, *Actinomyces scabies*, causes common scab of potato tubers, and makes the growing of potatoes unprofitable in some regions. The rough patches on the potatoes you buy at the store that look as if the potatoes had a skin disease are caused by this fungus, and a skin disease of the potato is what it is. Another, *Actinomyces bovis*, has already been met with in Chapter 9 of this book, as the cause of lumpy jaw in cattle. Most of the many kinds of Actinomyces that are prevalent in the soil, however, are harmless, both to man and beast.

Dr. Waksman is a soil microbiologist, one of a group of scientists of whom there are not many in the country. The field is not only difficult and rather involved, it is seemingly of little practical value. When a state legislature meets to appropriate moneys for the experiment station, the legislators are not likely to consider the efforts of a soil microbiologist worthy of much support. He is not among those hauled down to the capitol at appropriations time to impress the legislators with the benefits that the farmers are deriving from experiment-station research. He is endured, rather than encouraged. Which may be one reason why so little still is known about soil microbiology.

Dr. Waksman was a pioneer in the study of fungi and bacteria in the soil. During his decades of work on these soil microorganisms he had isolated and worked with many and various bacteria and fungi, among them species of Actinomyces and Streptomyces. Some of these produced substances toxic to certain other fungi and bacteria, and one of them, *Streptomyces griseus*, isolated from a manure pile, was found to produce an antibiotic poisonous to certain bacteria not inhibited or killed by penicillin. This is the species from which Dr. Waksman got streptomycin. The cultures used to produce this drug are, and remain, essentially Dr. Waksman's property. The industries which grow the fungus and produce the drug do so within the framework of a contract with him.

This is the usual practice with such molds, bacteria, and yeasts as are put to work commercially. Before an outcry is raised, however, against monopoly which works against the public benefit, it is well to face up to a few facts. No selfishness or churlish love of secrecy is involved. The only protection the discoverer of such a new biological process has is to conceal its exact nature. Commercially valuable molds, yeasts, or bacteria cannot be patented as can new mechanical inventions or even new varieties of roses, snapdragons, or apples, and unlike new publications and new comic strips, they are not protected by copyright laws. This situation exists in spite of the fact that one of the early plant patents taken out, after the laws were revised to permit the patenting of new plants, was for a new or different variety of cultivated mushroom. The spawn of this mushroom, which is the product sold to growers, is just mycelium. This mycelium, from any commercial standpoint, is not different from the mycelium of *Penicillium notatum*, from which penicillin is obtained, or that of *Streptomyces griseus*, from which is obtained streptomycin. Yet a new variety of the cultivated mushroom, *Agaricus campestris*, can be patented, but a new variety of *Streptomyces griseus* or *Penicillium notatum* or *Aspergillus niger* or any of the other molds exploited industrially cannot be. Why not, only a congressman can tell you. It is only to be expected that a firm or an institution which has spent years of time and large sums of money on developing certain productive strains of a mold or bacterium will feel that it should reap some of the benefits therefrom, even as others reap the benefits from patent or copyright protection. At the present time their sole protection is secrecy. This is a poor guarantee against piracy, but it is the only one they have.

The research worker delights in his job for its own sake, and is animated by a passion for finding out the causes of things. The rewards of the job are above all in the job itself. Yet this does not mean that he is immune to all other considerations and rewards, or incapable of sharing in other human motives — the desire for recognition, for approval, for support. In particular, he has good reason to wish that the nature of his work might be better understood by the general public than it appears to be. Not long ago there appeared in a book by a nationally read columnist some contemptuous waggery at the expense of the research worker, insofar as the fruits of the type of study the researcher engages in might be inferred from the titles of the Ph.D. dissertations in the Harvard Library. With all due respect to this columnist, he was hardly competent to express an intelligent opinion on even the titles of these theses, the only part of them that, by his own admission, he read.

His ridicule of research would scarcely be worth taking notice of if it were not for the fact that a good many citizens have the same idea. To them it seems that the research worker is living in an ivory tower, where he spends most of his time on abstruse hair splittings that have not the remotest connection with the sweat and struggle of everyday life. They think that every single research problem should answer a question of immediate practical importance, and do so in language the layman can understand. Should produce something that will improve the hard lot of the farmer, manufacturer, or mechanic right now. Today. The trouble, of course, is that many problems of immediate practical importance are not simple, as parts of the preceding account of the fungi will have made clear. Even to understand what the problem is we often have to push out into the unknown. And the ability even to frame the questions that are to be answered is the result not merely of the research worker's natural curiosity, but of knowledge increased as it is exercised, and of judgment tempered by experience.

Hundreds of problems have been and are today being solved by combined research such as that which converted a laboratory dream into an abundant supply of penicillin. Progress of this kind would not be possible if one requirement of every bit of research done was that both content and title should be immediately intelligible to every farmer, householder, radio commentator, or gossip columnist. Even with research, to be sure, we progress slowly, but in practical benefits alone we have advanced more rapidly since research has been incorporated into agriculture, medicine, industry, and even the social sciences, than we did before. The methods and attitudes of research may be one of the most important contributions Western civilization has made not just to the general welfare of the world but to world culture, in the sense of providing an environment more favorable not only to the physical but to the intellectual and spiritual growth of man.

Additional Reading

C. M. Christensen and H. H. Kaufmann, *Spoilage of Stored Grain*. Extension Folder 226, Agricultural Extension Service, University of Minnesota, 1964.

Robert D. Coghill and Roy S. Koch, "Penicillin, a Wartime Accomplishment." *Chemical and Engineering News*, 23:2310–2316 (1945).

G. W. Irving and H. T. Herrick, eds., *Antibiotics*. Brooklyn, N.Y.: Chemical Publishing Company, 1949. 273 pp.

D. Perlman, "Mycological Production of Citric Acid — the Submerged Culture Method." *Economic Botany*, 3:360–374 (1949).

George Smith, *An Introduction to Industrial Mycology*. London: Edward Arnold & Co., 1938. 302 pp.

Charles Thom and K. B. Raper, *Manual of the Aspergilli*, pp. 289–318. Baltimore: The Williams & Wilkins Co., 1945. 373 pp.

United States Department of Agriculture, *Mushroom Growing in the United States*. Farmer's Bulletin 1875. Washington, D.C., 1941.

CHAPTER II

Toxic Fungi

IN THE previous chapters you have been introduced to some of the ways in which fungi of various kinds affect us for good or ill. The present chapter aims to tell you something about a perhaps less pleasant or entertaining, but still in many ways interesting, aspect of their impact on us — that of toxicity, or poisoning. Some fungi, especially among the mushrooms, are poisonous in themselves; that is, some of the chemical substances in their makeup are toxic to us, and to some other animals, when eaten. It would be strange if this were not so, since all the larger groups of plants include species that are poisonous when eaten; the fungi are no exception. That certain mushrooms are poisonous when eaten has long been known. Relatively recently, however, it has been found that some of the fungi that commonly grow in a variety of foods and feeds may, under certain conditions, also produce substances toxic to man and his domestic animals. These various kinds of poisoning by mushrooms and their relatives, and by the fungi that grow in foods and feeds, are so common and important that anyone moderately interested and informed in biology should know something about them. Those who raise poultry or stock, or who provide feed for poultry or stock, should be vitally interested in toxicity of fungus-invaded feeds, since the success or failure of their enterprises may depend on recognition and avoidance of these troubles.

Poisonous Mushrooms

About 4000 species of mushrooms or gilled fungi have been described, and of these perhaps 30 or 40 species are known or suspected to be more or less toxic. The unsatisfyingly imprecise "suspected" and "more or less" simply indicate that our knowledge of the effects of some of these mushrooms when eaten is by no means as complete as might be desirable. There is good reason for this: There is no outward sign to indicate that a given kind of mushroom is poisonous when eaten. The only way this can be found out is for someone to eat it and be poisoned by it, and it is in this way that our knowledge of poisonous mushrooms has been acquired. It might be supposed that mushrooms of unknown edibility might be fed to laboratory animals to determine possible toxicity, but this does not necessarily work — for one reason, because what is poisonous to one animal may not be so to another. The mushroom *Amanita phalloides*, for example, is exceedingly toxic to man, and yet rabbits are said to be able to ingest at least a moderate amount of it without fatal consequences. One of the antidotes that have been tried to counteract poisoning in those who have eaten *A. phalloides* consists of feeding the victim ground up raw stomachs and brains of rabbits; I do not know of any authenticated cures of mushroom poisoning attributable to consumption of this revolting witches' brew.

Since there is no sign by which a poisonous mushroom can be recognized, there is no rule that enables those who pick and eat wild mushrooms to avoid the bad and select the good. There are, to be sure, a variety of tests or rules, rooted in folklore, that supposedly enable one to distinguish edible wild mushrooms from the unwholesome or poisonous kinds, but none of these rules or tests has any validity whatever. At times such rules seem to work because people without any technical knowledge of mushrooms pick and eat a great variety of wild mushrooms and almost never encounter any serious trouble. What saves them is that they may know by sight a couple of the most poisonous kinds, and avoid them. (They may, of course, avoid many entirely wholesome kinds too.) Also, the really deadly *Amanita phalloides* and the closely related variety or species *A. verna* usually are not too abundant, or are abundant for only a short time. *Amanita muscaria*, somewhat less deadly than *A. phalloides* and *A. verna*, is common and widespread, but it is so easily identifiable that once seen it can be recognized and avoided. Cases of fatal poisoning from eating wild mushrooms are relatively rare, fortunately, but not because of any "rules" which may be followed in gathering wild mushrooms to eat. A few cases will be described to illustrate the nature and kind of poisoning that may result from eating wild mushrooms, but first a few words about wild mushrooms in general.

Most of us want something more from our food than just that it does not poison us out of hand. Preferably it should be nourishing, and of attractive flavor and texture; and "nourishing" means not only that it contains some of the proteins, fats, carbohydrates, minerals, and vitamins that we need, but also that it is digestible. Many people believe that, nutritionally, mushrooms are closely akin to meat, and may even serve as a substitute for meat, instead of just a pleasing garnish for it. But probably few or none of the wild mushrooms have as much food value as the cultivated kind (which is grown on a very rich substrate, composted horse manure) and that has little enough, mostly a few minerals and some vitamin B. In food value wild mushrooms are much closer to rutabagas than to meat, and to be worth the trouble of seeking them out for food they should have at least an attractive flavor and texture. Some do, some don't.

Many kinds of edible wild mushrooms have a flavor of the material on which they are growing. *Armillaria mellea*, commonly known as the "honey cap," when growing on the stumps of recently cut oak trees is likely to have a strong enough flavor of tannic acid to pucker up your mouth, plus a certain sliminess that increases if it is boiled, plus a rather tough texture. It is perfectly edible, and many people gather it by the bushel, dry it, and eat it with avidity and enthusiasm all winter long. Others do not consider it to be suitable food for humans.

Some mushrooms, even when fresh and sound, have an odd odor and flavor, and some are essentially without either odor or flavor.

In texture the wild mushrooms range from fragile to firm to so leather-tough that no amount of cooking or chewing will reduce them to more than swallowable chunks. Mushroom enthusiasts nevertheless consume them with gusto.

Also, most wild mushrooms, by the time they are large enough to be seen and picked, already have been invaded by insect larvae. Small flies lay their eggs in the caps, or larvae invade the stems even before the mushrooms come above the ground. These larvae or maggots are small, and even when they are fairly numerous in the flesh of the cap and stem they may not be seen unless one looks for them carefully. They probably are relatively harmless, although as they tunnel through and consume the interior of the mushroom they deposit excreta, and large numbers of bacteria grow in that and in the tunnels — bacteria that may or may not be wholesome. To most people, wormy food is not acceptable, let alone appetizing, and if they were to find in the meat they bring home from the store the same number of maggots that are in the wild mushrooms they have gathered to cook with the meat, they probably would boycott the butcher.

We have, many times, picked seemingly sound, clean specimens of different species of wild mushrooms and put them immediately into insect-proof cages, so that no insects had access to them from the outside. Within two or three days the caps of all of these mushrooms were riddled with the tunnels of larvae, larvae that undoubtedly came from eggs laid on the mushrooms before they were picked. They were small larvae, to be sure.

Enough of that. The idea is that many wild mushrooms, even if edible, may not be attractive in flavor or texture, and that they may consist of something more than just mushroom. Now to a few of the poisonous kinds.

Amanita phalloides and *A. verna*. These closely related varieties or species are equally poisonous. O. E. Fischer, who wrote the section on poisonous mushrooms in Kauffman's classic *Agaricaceae of Michigan*, states that it would be conservative to estimate that nine-tenths of the deaths from poisonous mushrooms have been due to *A. phalloides* (and *A. verna*). According to him, in 1900 in France

100 people died from eating *A. phalloides*, and, in 1912, 153 deaths were ascribed to eating this mushroom in France. Even a small portion is sufficient to cause fatal poisoning. One child died after consuming one-third of a small cap of the fungus, and two children died after eating a small piece of bread soaked in the juice of the mushroom.

Symptoms of poisoning do not appear until 8 to 15 or 20 hours after the mushrooms have been eaten, and by then the victim is likely to be beyond help, except for such medication as may ease the suffering. Among the symptoms of poisoning for *A. phalloides* are intense pains throughout the body, vomiting, diarrhea in which greenish liquid, blood, and mucus may be voided, cramps, convulsions, jaundice, delirium, and coma, followed by death. And death comes not quickly, but only after 8 to 20 days of horrible suffering, by which time some of the internal organs of the victim may have undergone extensive degeneration. Mortality is said to range from 60 to 100 per cent, and those who recover (probably because they ingested only a very small amount of the fungus) may not regain full use of their limbs for some weeks or months. Truly a mountain of misery for the minor pleasure of having eaten a small portion of a wild mushroom.

Amanita phalloides is said to contain three poisons, phallin, phalloidin, and amanitin, the latter with the formula $C_{43}H_{45}O_{12}N_7S$ and with a molecular weight of 609. A quantity of 5 micrograms (a microgram is a thousandth of a milligram, or a millionth of a gram) or about 1/6,000,000 of an ounce, injected into a mouse was fatal to it. A potent poison indeed.

Amanita muscaria. This large and showy mushroom is somewhat less toxic than its near relative *A. phalloides*, but eaten in sufficient quantity it is lethal. In lesser quantities it causes an intoxication akin to but more brutal than that caused by alcohol. Up to within at least very recent times it was, and may still be, consumed by the natives of some regions of Siberia to produce a hallucinatory state in which the mind melds with the universe and the future seems to become clear. There appears to be a rather narrow margin of safety between the amount needed to produce the desired state of euphoria, with its attendant revelations, and the amount that causes severe illness or even death. One writer states that the natives in question would give furs worth $20 for a single dried specimen of *A. muscaria*, although as many as ten specimens might be needed for a really rousing drunk. Much of the toxin is excreted within a short time in the urine, and it is reported that this would be consumed among these people to continue the party — it was said that it could be passed in this way through as many as five persons before losing its potency. Reality must have been pretty brutal to warrant an escape of this sort.

Lepiota morgani. The genus *Lepiota* contains a number of choice edible mushrooms, but one species, *L. morgani*, is poisonous, causing severe illness but seldom resulting in death. It can be distinguished from some of its close relatives only by the fact that it produces pale green spores, while the edible species of *Lepiota* produce white or pale tan spores. In this case a spore print is absolutely essential to correct identification, and those who identify wild mushrooms by guess, or by majority vote of the assembled club members, may expose themselves to fairly unpleasant poisoning. Strangely enough, some people appear to be able to eat *L. morgani* without any ill effects whatsoever, and even relish it.

Coprinus atramentarius and *C. micaceus*. These two common "inky caps" are edible without any question, and even delicious, of excellent flavor and delicate texture. They might even have some food value. However, there seems to be good evidence that if alcohol is consumed after they are eaten, poisoning may result. Poisoning of a not too severe sort, and of relatively short duration, but poisoning nevertheless. One account tells of a man who ate six pounds (good heavens!) of *Coprinus atramentarius* for dinner, washed down with several bottles of wine, and who soon thereafter was taken with palpitation of the heart, shivers, and reddening of the body. It was not stated what he engulfed at the meal besides the six pounds of mushrooms and half a gallon or more of wine, but presumably these were accompanied by other assorted delicacies in suitable quantity. No wonder he shivered and turned red, mushrooms or no mushrooms.

Gyromitra esculenta. Commonly known as the "false morel," this mushroom comes up in abundance in the spring throughout much of the north temperate zone around the world. It has a strange record. By some it is rated as one of the best of the edible fungi, and is eaten regularly and in quantity without any ill effects whatever. Yet there is no question but that, at times, it has caused severe illness, and even death, of those who ate it. At one time it was licensed to be sold in the public markets of most cities of Europe where wild mushrooms are gathered for sale, but now it is banned in most or all of them. No one knows why it sometimes is poisonous, sometimes not.

It has been suggested that it contains a water soluble acid that is leached out if it is boiled, and so if it is cooked in this fashion, and the water thrown away, the mushroom is not poisonous. Another suggestion is that there are varieties of it, some edible, some poisonous. A third suggestion is that the poison is a product of partial decay of the mushroom by bacteria, and that really sound specimens are wholesome. In all that we have tested, bacteria are present in the fresh and apparently sound specimens, and even accompany the ascospores shot out from the asci, and so there may not be *any* fruit bodies of this fungus that have not been invaded by bacteria to some extent. Whatever may be the right explanation of its occasional poisonousness, there is no doubt that eating *Gyromitra esculenta* is accompanied by some hazard.

The above accounts of some of the poisonous mushrooms should be sufficient to indicate that wild mushrooms should not be eaten indiscriminately by anyone who wishes to lead a long and happy life. If wild mushrooms are gathered for eating, two simple rules should be followed: (1) Learn thoroughly the characteristics of the kinds you pick, so you can identify them with 100 per cent certainty. (2) Pick only those that you know are good and wholesome.

The study of wild mushrooms can be a lot of fun in itself, quite aside from their culinary interest; you don't have to eat them to enjoy them, any more than you have to eat wild flowers or wild birds to enjoy them.

Hallucinogenic Fungi

A number of compounds related to and containing the chemical substance indole as part of their structure are known to affect the nervous system of man in various ways, sometimes heightening perception, releasing pent-up emotions, eliminating conflicts, and producing hallucinations of various sorts. The most common of these hallucinogenic drugs are mescaline, present in the peyote cactus, *Lophophora williamsii*; psilocybin and psilocin, present in several species of mushrooms such as *Psilocybe mexicana* and *Stropharia* (or *Psilocybe*) *cubensis*; and d-lysergic acid diethylamide (LSD), present in the ergot fungus, *Claviceps purpurea*, and produced or elaborated by some other fungi also. The peyote cactus of the southwestern United States and adjacent northern Mexico, and the hallucinogenic species of *Psilocybe* in central and southern Mexico, have for thousands of years been eaten by the natives of those regions, sometimes as a sort of community ritual, sometimes by the medicine men or women to get in tune with the infinite.

One recent account of the uses and effects of *Psilocybe* among the natives of southern Mexico makes these mushrooms seem really remarkable. According to this, the natives fear and adore these mushrooms, and regard them as the key to communication with the Deity. Traffic in and use of the mushrooms are secret, and they can be conversed about only in whispers among trusted friends, behind closed doors, and in the dead of night. Partaking of the mushrooms enables the medicine man or woman to predict the future (anyone can predict the future — the trick is in hitting it right; even the weather bureau knows that), locate lost or stolen property, and communicate with far-distant friends and relatives.

Tests by physiologists, psychiatrists, and other interested members of the medical profession in the United States, in which these drugs have been administered to volunteers, indicate that the drugs do indeed affect the mind in various ways, as has long been known. But whether the candidates to whom these drugs are administered really get in tune with universal truth or merely get out of tune with everyday reality seems to be still in question. These drugs are of interest to investigators in the medical sciences precisely because they do alter some aspects of consciousness; the drugs or their derivatives may serve as a tool for learning more about the chemistry of consciousness and the chemistry of normal and abnormal psychology, and may be of value in the treatment of some mental ills, including schizophrenia. They do not seem to have worked any special wonders so far, even in the lost and found department, but study of them continues.

Mycotoxicoses

Mycotoxicoses means diseases caused by fungus toxins or poisons, and it is the general term applied to diseases resulting from eating foods or feeds containing toxins produced by fungi that have grown in these materials. We and our domestic animals probably always have been subjected to a certain amount of this sort of poisoning, but only recently have mycotoxicoses come to be recognized as a real and present danger. If we still are a long way from being able to judge accurately the seriousness of the problem of mycotoxicoses, at least we now are aware that such a problem exists, and that in itself is a great advance over the state of our knowledge, as well as our state of mind, so short a time ago as 1950 or even 1955.

In the 1880's and 1890's several cases of poisoning of farm horses in different places in the United States were attributed to the animals having eaten corn invaded by fungi. The evidence was not clear-cut, as so often is true of cases of this kind, and nothing much was made of these reports. However, a few horses that were deliberately fed with damaged corn from suspected lots of feed died. No large numbers of test animals were involved, partly because few research budgets in those days, and few enough even now, can endure the expense of buying, housing, caring for, and feeding 20 or 30 horses for a given test of suspected toxic feed, with the probability that the tests will have to be repeated each year for at least several years to get really convincing results. Horses were relatively expensive even in those good old days before our currency became converted into what sometimes seems to be play money. However, the early students of this problem obtained at least good circumstantial evidence that something in the damaged corn was toxic enough so that when fed to horses for six weeks or so it killed them.

In 1934, 5000 horses were reported by a reliable veterinarian to have died in central Illinois alone, supposedly from consuming cornstalks or corn invaded by fungi. It seems fairly conservative to estimate a value of $200 to $300 for a farm horse in Illinois then, since undoubtedly many of these horses were purebred draft animals. This would mean a loss of 1 to 1.5 million dollars in central Illinois alone in that year. Yet, so far as can be determined, no one became concerned enough about the problem to follow it up. Losses of lesser magnitude were reported by veterinarians in Iowa and other midwestern states, about the same time or slightly later, from consumption of corn invaded by fungi. The disease was called "cornstalk disease" or "moldy corn disease" and, from the symptoms displayed in the affected horses, was known as "staggers" or "blind staggers." Some of the postmortem findings in the horses were similar to those encountered in encephalomyelitis, caused by a virus. However, some of the horses that died of this disease in Illinois in 1934 were checked for possible pathogenic bacteria and viruses, the tests including inoculation of brain tissue from recently expired horses into other animals. No disease resulted. This was an added reason for suspecting something in the corn itself as the cause. A veterinarian in Iowa in 1937, in fact, suggested that "close cooperation with plant mycologists would be desirable." This was a very enlightened view, but again nothing much came of it.

Critical and productive research on mycotoxicoses in the United States got under way in the 1950's, when Dr. Forgacs, a competent mycologist and microbiologist, began to work with veterinarians on a number of animal diseases with hitherto unexplained causes. Within a few years he and his coworkers proved that several common, widespread, and important diseases of animals were caused by consumption of feed that had been invaded by fungi. For one reason or another their results did not cause the stir that might have been expected. In 1957 they reported that one of nine strains of *Aspergillus flavus* which Forgacs had isolated from damaged corn, suspected of poisoning pigs, would, when inoculated into moist autoclaved corn and incubated for a month, kill pigs to which it was fed.

Note that only one of nine isolates of the fungus was toxic; had they tested only one or a few isolates they might well have missed discovering that grain invaded by *A. flavus* can be toxic when eaten.

In 1960, 100,000 turkey poults in England died of mysterious causes. The company which had manufactured the feed with which these turkeys had been fed had to make restitution to the growers, at a cost of $500,000. This was, however, considerably less than the monetary loss suffered from the death of horses in central Illinois in 1934, especially considering that a dollar in 1934 was worth about three times as much as it was in 1960.

The workers in England undertook to find the cause of the turkey-X disease, as it then was known. They explored the possibility that it might be caused by bacteria, by viruses, and by chemical poisons in the feed — all without result. Finally the toxic feed was traced back to one lot of peanuts from Brazil, and as a last resort the possibility of a fungus toxin was considered. The lot of peanuts was found to be infected with *Aspergillus flavus*, and, before long, it was found that some strains of this fungus, when growing in peanuts, elaborated several potent toxins, and one or more of these were proven to be responsible for the death of the turkeys.

This, plus additional work of others, and several reviews by Forgacs of known mycotoxicoses, focused attention on the problem of feeds and foods made toxic by fungi, and now research on this is under way in a number of places. From the evidence to date it seems probable that we are only on the threshold of a whole complex of similar problems. Accounts of a few of the diseases caused by eating feeds invaded by fungi will serve to illustrate what is involved. More detailed information about them can be found in the papers cited at the end of the chapter.

Fusarium. The genus *Fusarium* includes many species, and several of these species are common in grains such as wheat, barley, oats, sorghum, and corn, all of which are widely used as animal feeds and human foods. Some of these species invade the kernels of the plants during the growing season and cause discoloration and shriveling of the kernels. When this is obvious the grain is said to be "blighted." Barley blighted by *Fusarium* may be toxic to man and to pigs, as was shown more than 30 years ago. In one test a water extract of a culture of *Fusarium* isolated from blighted barley was administered to a hundred-pound pig via a stomach tube, and caused the pig to vomit eight times in half an hour. When grain inoculated and heavily overgrown with this fungus was offered to pigs, they refused it; after consuming a small amount they recognized it as unwholesome, and would starve rather than eat more of it. Tests in a number of places, in which corn or other grain made toxic by growth of *Fusarium* on it was refused by pigs, gave rise to the belief that pigs could safely be allowed to discriminate between grain that was wholesome to them and grain that was not.

There is good reason to question this belief, since pigs frequently do consume enough feed made toxic by the growth of *Fusarium* or other fungi on it to make them ill, or even kill them. The toxic compounds produced by a given strain of *Fusarium* may not be the same as the compounds that make the grain unpalatable to the pigs, and some strains of some fungi evidently do not in any way disclose their toxic properties to even the smartest of pigs. There may be a whole host of complexities involved, but in any case pigs and other animals do consume feeds made toxic by fungi, and in sufficient amounts to cause illness and death. If pelleted or otherwise processed feeds are made in part from grains heavily invaded by fungi that cause toxicoses, any bad flavor contributed by the fungi may be concealed by other ingredients of the feed. Lots of pelleted feeds have been obtained that when fed to experimental animals were consumed with no evidence of aversion and yet caused serious illness or death in the animals to which they were fed. Animals definitely cannot be depended upon to reject feed that is unwholesome to them.

A disease called toxic alimentary aleukia (which means a disease characterized by a decreased number of leucocytes or white blood cells, and caused by something toxic that was eaten) was widespread in the early 1940's among people of the Orenburg district of the U.S.S.R. People, in this case, not pigs. The persons in whom this disease appeared had subsisted in part upon grain, especially proso millet, that had been left over winter in the fields, and so had been subjected to invasion by various fungi. After a good deal of work with different fungi isolated from the toxic grain, the major blame for the poisoning was placed on the fungus *Fusarium sporotrichioides*. The temperature prevailing when the fungus was growing in the grain greatly influenced the production of toxin, and only after grain invaded by *F. sporotrichioides* was subjected to alternate freezing and thawing in the late winter or early spring did it become highly toxic. Some strains of the fungus, when grown under conditions that favored production of toxin, produced much more toxin than others.

These two factors — the strain of the fungus and the temperature at which it invades the substrate on which it is growing — are not the only things that influence toxin production, but they are important. A given fungus can be harmless or lethal, depending on the strain of the fungus, the conditions (especially temperature) under which it is grown, the length of time it is incubated on inoculated grain, and whether it is growing alone, as a pure culture, or in a mixture with other fungi. This complicates the situation. Some investigators have fed grains invaded by fungi to various experimental animals and found no ill effects on the health of the animals. From this they concluded that fungi in feeds were harmless. All they really found was that the strains of the fungi they were working with, grown under the conditions maintained in their tests, were not observably toxic. The important thing to determine, of course, is under what conditions what fungi will produce materials toxic to whatever or whomever consumes that food or feed.

It might seem that the combination of fungi and circumstances required to produce markedly toxic feed would be so complex as to seldom occur in nature, but this definitely is not so. The seemingly intricate set of circumstances that result in the production of toxins may be encountered rather commonly in nature. The investigator's problem is to determine what these factors are, and to work out the details of their interplay, a task that will require years of work by men well trained in veterinary pathology, animal and fungus physiology, and mycology.

The "estrogenic syndrome" in swine involves development of a swollen, edematous vulva in female pigs, shrunken testes in young males, enlarged mammary glands in the young of both sexes, and abortion in pregnant sows. It is a disease of considerable importance in some of the swine-growing areas of the United States, and until recently the cause was a mystery. The evidence now is conclusive that it is caused by eating grains, especially corn, invaded by certain strains of *Fusarium graminearum* (the perfect or ascospore stage of which is known as *Gibberella zeae*). This fungus is not common on corn at harvest, but once the corn is stored on the cob in a crib (which throughout the Corn Belt is the common way to store corn that is to be fed to livestock) and exposed to the weather, it may be invaded by this and a multitude of other fungi. As with *F. sporotrichioides*, a period of low temperature, or alternating moderate and low temperature, is necessary for the production of toxin. Toxin production also is influenced by the length of time the fungus is grown in the corn, reaching a peak after a few weeks, then rapidly declining. Toxin production also is influenced by other fungi associated with *Fusarium*.

A large producer of pelleted feeds found that one lot of their feed, fed to their own experimental herd of swine, produced the typical estrogenic syndrome outlined above. When fed to white rats and guinea pigs for only 7 to 10 days in our tests, it produced greatly enlarged uteri in these animals, one of the standard symptoms by which estrogens in feed can be recognized. Chemical tests showed that this feed contained a fairly large amount of the same estrogenic compound that was produced by a strain of *Fusarium* known to cause the estrogenic syndrome in swine. There is almost no question but that one or more of the grain ingredients of this feed had been heavily invaded by an estrogen-producing *Fusarium*. What disposition was made of the remainder of this toxic batch of feed? If anyone knows, he is not telling.

Are there still other species or strains of *Fusarium* that, growing under the right conditions, produce toxic substances that cause diseases other than those found so far? We do not know, but it seems highly probable.

Stachybotrys. Outbreaks of an unusual disease of horses in the Ukraine in the 1930's led to an intensive search for the cause. As usual, mycotoxins were considered only after all other possibilities had been exhausted. Within two years *Stachybotrys atra,* a fungus common on straw, hay, and cellulose substrates in general, was found to be the cause. When small amounts of hay or straw overgrown with this fungus are consumed regularly, a chronic type of the disease results, but when larger amounts of materials invaded by the fungus are eaten at one feeding, the disease may appear within six hours, and result in death in less than 24 hours. Not only horses, but also cattle, sheep, swine and other kinds of domestic animals may be affected. In one test, a culture of *S. atra* grown in a petri dish was fed to an 800-pound horse, and resulted in death of the horse within hours. As would be expected, not all strains of *S. atra* are toxic.

Intensive work was undertaken with this and other mycotoxicoses in the U.S.S.R. in the 1930's, and hundreds of research papers on these diseases were published there before 1940, while we in the United States had got no farther than a suggestion by a veterinarian that "close cooperation with plant mycologists would be desirable." A central research institute for the study of these diseases was established in the U.S.S.R., with a number of branch stations devoted to study of mycotoxicoses, and courses dealing with mycotoxicoses were organized in colleges of veterinary medicine. We are a long ways behind the U.S.S.R. in this work.

Aspergillus and *Penicillium.* Several species of *Aspergillus* are common and sometimes abundant in grains and feeds, and in some foods, including wheat flour milled in the United States, Canada, and Mexico, almost the only countries where flours have been tested for the presence of these fungi. *Aspergillus flavus,* as mentioned above, was found to produce a toxin that was lethal to turkeys in England. It since has been found to be lethal to some other kinds of poultry also, especially ducklings. The same fungus was found to be among the "numerically dominant" fungi in wheat flour from mills in Canada, and is not uncommon in the flour from some mills in the United States. Is flour wholesome when contaminated with spores of *A. flavus*? We do not know, but it would seem desirable to find out. Known amounts of the toxin produced by *A. flavus* were fed to mice for some weeks, and had no observable deleterious effects on their health, but if some men are mice, mice still are not men.

The business of moldy peanuts turned out to be more than just peanuts. Peanuts are eaten in various forms by man, including peanuts as such, peanut butter, and peanut flour. Toxins produced by *A. flavus* have been found in peanuts stored in warehouses in the United States, which was to be expected, since toxin-producing strains of the fungus are by no means rare. There is no reason to cry havoc, since both the Food and Drug Administration and the peanut processors are aware of the problem and of the possible health hazards involved, but there is very good reason for constant vigilance to assure us that the peanuts and peanut products we eat are wholesome.

Aspergillus flavus grows in many things besides peanuts, including stored cottonseeds, grains of all kinds, and feeds. A lot of pelleted poultry feed sent to us from Mexico for examination, because it had caused illness in broiler chicks to which it had been fed by the producers of the feed, killed every chick to which it was fed in our tests. The symptoms suggested aflatoxin, and the feed when cultured was found to carry a rather heavy load of *A. flavus*.

Some batches of feed are made from ingredients invaded by fungi, materials about whose wholesomeness there may be some question, to put it mildly. This is a problem that the feed-producing industries and some of the food processing industries are going to have to face. From the standpoint of our health and that of our animals, the sooner intensive studies of this are undertaken, the better.

Other species of *Aspergillus*, such as *A. chevalieri* and *A. glaucus*, have been found to produce toxic materials when growing in feeds. We have encountered batches of prepared feeds that were so heavily invaded by these and by related fungi that the feeds were heating and losing weight; the manufacturer of this feed was concerned only about the loss in weight resulting from the heavy invasion of the feed by fungi, not about any possible toxicity.

One or more species of *Penicillium* have been shown to cause what is known as "hemorrhagic syndrome" in chicks, characterized by severe internal bleeding and not uncommonly resulting in death of the affected chicks. *Penicillium* is exceedingly abundant in some feeds, and we need to know more about the circumstances under which it may produce toxins.

Foods that are relatively free of fungi when bought may be invaded by fungi in the home. Jams and jellies and preserves, fruits and vegetables, all kinds of flour, bread and biscuits and other bakery products, all are subject to fairly rapid invasion by fungi. If these products become obviously moldy or develop a musty taste, they are likely to be discarded, although there may be a considerable development of fungi in them before it becomes obvious to sight or taste. It seems unlikely that this poses a serious health problem, aside from the respiratory allergy that may be induced by inhaling their spores, but the point is that, at present, we just do not know. Rather than close our eyes and our minds to the possibility of slow poisoning from foods invaded by fungi, we ought to find out whether such fungus-invaded foods are deleterious to our health.

Summary

Some wild mushrooms are poisonous, a few of them extremely so, and no reasonable person should risk eating any wild mushroom of whose identity and edibility he is not absolutely certain. The consequences can be so horrid that the risk involved just is not worth while.

The problem of foods and feeds made toxic by fungi growing in them is relatively new. It has been investigated intensively and extensively in U.S.S.R. since about 1930, but only since about 1955 has work with mycotoxicoses been under way in the United States. About a half dozen different diseases of poultry and other domestic animals, caused by invasion of feeds by fungi, have been described, but unquestionably others remain to be discovered. Occasional batches of pelleted or otherwise processed feeds have been found to contain sufficient mycotoxins to cause serious illness and even death of the animals that consumed them, and the problem may be larger than anyone now suspects.

The production of toxins by fungi growing in foods and feeds is influenced by many factors — the strain of the fungus, the conditions under which it is growing, the length of time it grows in a given material, the associated fungi, and the nature of the animals that consume it. Much more work will be required, by teams of investigators expert in diagnosis of animal diseases, animal and fungus physiology, and mycology, before we can even begin to get the answers we need to evaluate the role of mycotoxicoses in our own health and well-being, and in the health and well-being of domestic and wild animals.

Additional Reading

J. Forgacs, "Mycotoxicoses — the Neglected Diseases." *Feedstuffs*, 34:124–134 (1962).

J. Forgacs and W. T. Carll. "Mycotoxicoses." *Advances in Veterinary Science*, 7:273–382 (1962).

M. A. Spensley, "Aflatoxin, the Active Principle in Turkey 'X' Disease." *Endeavor*, 21:582 (1963).

CHAPTER 12

Experiments with Fungi

STUDENTS and teachers in biology classes in high schools and colleges have become increasingly aware of and interested in fungi during the past decade or so, and requests have been received by various departments in many universities for demonstration materials or experiments that would aid them in learning more about the nature and habits of fungi. It is not at all difficult to grow some of the common and uncommon and striking fungi in a fungarium such as that described below. Also only a moderate cash outlay is required for materials and equipment for making cultures of fungi as the professionals do it — agar and a few other ingredients of culture media, a pressure cooker for sterilizing the media, and petri dishes in which to make the cultures. All the procedures described below have been used repeatedly at the University of Minnesota and elsewhere in beginning classes in plant pathology, forest pathology, and mycology, and they bring the fungi to students simply and yet often dramatically. (Plate 14 shows some of the ingredients and equipment needed.)

Rots of Fruits and Vegetables

In almost any grocery store you can find a few oranges in various stages of decay, and usually the decayed portion will be covered with a powdery green mass of spores of Penicillium, the most common cause of this rot. Get one or two of these moldy oranges, and a number of sound ones. Rub a common pin or needle or toothpick in the spores on the surface of the decaying orange, then stick it through the skin of a healthy one. By this means you inoculate the sound orange with the spores of the fungus. Attach a small tag to the pin or needle or toothpick, and write on the tag the date on which the inoculation was made. Then put the inoculated orange, and a sound one that you have not inoculated, in a small plastic bag, fold the end of the bag over, and put it where you can keep an eye on it.

Ordinarily within a few days a discolored spot appears on the skin of the inoculated orange, around the point of inoculation. This spot enlarges fairly rapidly from day to day, and in a week or less a heavy crop of spores of Penicillium is produced on the decayed portion. Eventually the entire fruit will be decayed, and the decay will spread to the orange that was not inoculated. You can dust some of the spores onto agar in a petri dish, described below, and find out whether you have, on the orange, just one fungus alone, or whether others are present. Once you have a pure culture of the fungus on agar, you can take spores from the culture and inoculate oranges, and see whether they decay. If they do, you have proved that a pure culture of this Penicillium causes decay of oranges.

A simple experiment, to be sure, but had you done it in the early 1800's you would have been way ahead of the biological thinking of the times, and probably would have rated as a pioneer in plant pathology. Micheli, an Italian botanist, grew some common fungi, including Penicillium, on freshly cut pieces of melon rind way back in 1527, and so proved that these fungi were living, growing plants, which was a new idea at the time. It still required over 300 years more for anyone to get the idea that these fungi might cause rots and other diseases of plants. The rots and the fungi were there — but the idea was not.

The steps you have just gone through in the simple experiment outlined above were formalized in the 1880's as "Rules of Proof," usually called Koch's Rules of Proof because Koch, a famous German bacteriologist, established them. He said, essentially, that to prove that any specific fungus or bacterium causes a specific disease, it is necessary to: (1) Find the fungus or bacterium in constant and regular association with the disease. (2) Isolate the organism and grow it in pure culture. The idea of this is to make as sure as you can that what appears to be a single fungus or other microbe actually is a single one, and not a mixture of two or several, with one causing the disease and another or others accompanying or following it. (3) Inoculate sound specimens with pure cultures of the organism and cause typical symptoms of the disease on the inoculated specimen.

If you go through the above steps several times with the fungus that causes the decay of oranges you can be fairly confident that the Penicillium you have isolated causes the decay. In this particular case there is very little room for doubt, because the Penicillium almost always occurs on the orange in pure culture. By going through these simple steps you have acquired something more than just the bare fact that this fungus causes a rot of oranges; you have also established a principle of procedure that can be and must be followed in the study of many kinds of diseases of plants and animals. You also have acquired the rudiments of the scientific or experimental approach. Given time and opportunity this approach might even diffuse into your everyday thinking and make you more skeptical of statements that are not supported by experimental proof. You will not then necessarily be educated, but at least you will have acquired one of the necessary attributes of an educated person.

If you want to make the experiment a bit more complex from here on, you can use spores from the orange or from your pure culture on agar to inoculate lemons, grapefruit, apples, potatoes, or other fruits and vegetables. In this way you will find out whether the Penicillium that causes decay of oranges is a specialist that attacks only oranges, or whether it can attack citrus fruits in general, or many kinds of fruits and vegetables. You also can inoculate a half dozen oranges, put two in each of three plastic bags, put one set in a warm room, another in a cool room, and another in the refrigerator, and so determine the effect of temperature on the rate of development of the decay.

There are a multitude of rots of fruits and vegetables caused by different fungi, and many of these are illustrated in the United States Department of Agriculture bulletins listed at the end of this chapter. Normally it is possible to get an assortment of these rots in any wholesale fruit or vegetable market, so that you can work with different ones. If the specimen you pick out is in the last stages of decay it may be overgrown by a number of different fungi and bacteria, and inoculation of a sound fruit or vegetable with spores from the surface of the decayed one might not result in rot. In that case, take a small piece of the decayed flesh of the infected specimen and put it under the skin of a fresh, sound one. The fungus responsible for the decay usually will spread out into the inoculated flesh before the secondary fungi. Transfer a bit of this fresh decay to another sound specimen. With a few such successive transfers you ordinarily can recover in pure form the fungus that causes the rot. Then you can culture the fungus on agar and inoculate it into a number of other fruits and vegetables, and put these at various temperatures to see what the fungus will do. If you have available a botany or biology text that gives a good account of the fungi, you may even be able to identify some of the fungi.

Damping Off

Seedling diseases are described briefly on pages 94–95 of this book, and the account includes an experiment that serves to illustrate them, but perhaps it is just as well to enlarge on this and give more explicit and detailed directions. The general idea is that germinating seeds and tender young seedlings of many kinds of plants are very susceptible to invasion by a variety of fungi that inhabit the soil. These diseases are likely to be more prevalent in moist soil and wet weather, whence the name "damping off." Before the days of seed-treating fungicides and other means of control, damping off frequently was a plague in home gardens, truck gardens, greenhouses, forest tree nurseries, and other places where seedlings are grown in large numbers. If you make the following experiments you can easily see why.

You will need a dozen or so clay pots of about four-inch top diameter, some black garden soil, sand, and seeds of various kinds. We have used seeds of tomatoes, radishes, lettuce, beets, carrots, alfalfa, zinnias, marigolds, pine, and spruce, but seeds of any kind will serve so long as they are fairly small and have a high germination percentage. If you buy these seeds, get those that have *not* been treated with a fungicide to protect them from damping off — the label on the packet or box of seeds will tell you if a fungicide has been applied to the seeds.

Fill some of the pots to within an inch of the top with sand, some with black soil, pack it down, soak well with water, and level the surface. Count out as many sets of 25 seeds each as you have pots. Plant one set of 25 of each kind of seed in sand, another set of 25 in black soil, cover with sand the seeds planted in sand, and cover with soil those planted in soil, to a depth of ¼ inch. If the air in the room where they are to be kept is dry, cover the pots with a piece of glass or thin plastic. A low temperature, in the range of 60–70 degrees Fahrenheit, will result in slower germination of the seeds, slower development of the seedlings, and more damping off, than a temperature in the 80's, so put the pots in a place where they will be at least moderately cool at night.

To determine the germination percentage of the seeds you use, count out a hundred of each kind, put them on a piece of moist paper toweling, cover them with another moist paper towel, and wrap this loosely in a sheet of waxed paper or aluminum foil to keep it from drying out. The paper towel on which the seeds are placed, and that used to cover them, should not be sopping wet so that the seeds are submerged in water, just wet enough so that the seeds can take up enough water to germinate. Uncover the seeds every day to see how they are getting along. Count and record the number that germinate — with high-quality seeds of most kinds of plants this should be 90 per cent or more.

Also record the number of seedlings that come up in sand and in black soil in the pots where you have planted the seeds. Usually the number of seedlings that appear in black soil will be lower than the number that come up in sand. Fungi that cause damping off are likely to be much more numerous in black soil than in sand, because in black soil they have much more plant remains to feed on, which is what makes the black soil black. The percentage of seedlings that appear in sand may be lower than the percentage germination on moist paper, since some damping off fungi thrive in sand, too. In many forest tree nurseries, where the seedlings of our future forests are grown, the seed beds have long been made of sand, but even so damping off may eliminate a large percentage of the seedlings if control measures are not used.

Usually some of the seedlings that come up in the black soil will be attacked within a few days at or just below the ground line by these same damping-off fungi, and will topple over. If only a few seedlings have been killed in the black soil, plant another set and see whether the number killed does not increase; ordinarily it does, because the fungi that cause damping off have become established in the soil. After several crops of seedlings have been grown in such soil the damping off fungi may be so abundant that few seedlings will survive long enough to even break through the soil. Some of these dead and decayed seedlings can be seen by digging up the soil, but they decay so rapidly that after a few days even their remains cannot be found.

Once you have soil in which these damping-off fungi are established — and if you use garden soil in the first place damping off is likely to be severe in the first crop of seedlings — you can make various other experiments. Put a half teaspoonful of this soil on one side of a pot of sand, at the time you plant the seeds, and see if the disease progresses from the side of the pot you inoculated to the other side. Heat several pots of black soil, loaded with damping-off fungi, in the oven for half an hour, and after the soil is cool, water it and plant in it seeds of a kind susceptible to damping off. Heating the soil of nursery beds with steam is one method of controlling damping off. Out of doors the control by heating is only partial and temporary, because the steam does not penetrate far into the soil and so eliminates fungi from the

surface layers only. The fungi grow into this soil again from below and from the sides, and if the first fungi that colonize the
steamed soil happen to be those that cause damping off, the disease hazard within a short time may even be increased.

You can buy, in dust form, fungicides that have been developed to control damping off, and test their effectiveness. Count
out the seeds you wish to treat, put them in an envelope, add a
small amount of fungicide, and shake the seeds vigorously a few
times, so that all of them become coated with the dust. Then
plant them in soil in which damping off has previously occurred.
As a comparison, plant an equal number that have not been
treated. These seed-treating fungicides are poisonous, as the labels
on the containers will tell you, and they are not to be inhaled
or swallowed or scattered around indiscriminately.

Cultures of Fungi

As stated above, about 400 years ago an Italian by the name
of Micheli grew some of the common molds in more or less pure
culture on freshly cut slices of melon flesh. You can grow many
kinds of fungi on slices of fruits and vegetables too, and on slices
of bread. For this, home-baked bread is preferable to bakery
bread, because most bakery bread has mold inhibitors added to it
to keep molds from growing too vigorously on it. These inhibitors are fairly gentle in action, however, and usually will not
keep bread free of molds for more than a few days. In all our
tests in which bread has been used for growing molds, the molds
have grown vigorously within a week, and so probably most any
freshly baked bread will serve the purpose moderately well.

Take a sliced loaf of bread out of its wrapper, separate some
of the slices, scatter over them ground pepper from your pepper
shaker or from a can of ground pepper fresh from the store, put
the slices back together as a loaf, put this back in the wrapper,
and put it in a plastic bag so the bread will not dry out too rapidly. Keep for comparison a similar loaf that has not been opened.
Put both in a warm place, leave them for a week, then open them
and see which has the greater variety of fungi growing through

the bread. Instead of or in addition to inoculating the bread with pepper, just open the loaf and leave the slices exposed to the air of the home for half an hour, then put them back together, put the loaf back in the wrapper, and put this in a plastic bag and leave it for a week. Or scatter some dust from the vacuum cleaner, or soil, on the opened slices, then put them back together and incubate as you did the others. A different complex of fungi is likely to grow out from each of these materials, and many of the colonies will be so distinct that you can identify them.

It is not at all difficult, and for many purposes it is much more satisfactory, to grow the fungi on an agar medium in culture dishes, as the professional students of fungi do. Various sorts of culture dishes, tubes, and bottles are used for this, but for much of the work of sampling different sorts of materials for fungi petri dishes are preferable. They offer a large surface on which many colonies of fungi can grow at the same time, and when the colonies have developed to the stage where they are producing spores, the covers of the dishes can be removed, the opened dish put on the stage of a microscope, and the colonies observed right in place.

Petri dishes come in various sizes, but the standard size for most culture work is 90 millimeters wide and 15 millimeters deep. Those of pyrex glass cost, at 1961 prices, about 80 cents each, while those of plastic, sterile when received, cost from 6 to 10 cents each, depending on the number bought. The pyrex dishes can be used over and over again, can be sterilized in a pressure cooker or by heating in the oven at home, and are more transparent so that young colonies can be observed from below, as sometimes is desirable, without opening the dish; but for some purposes the plastic dishes are equally satisfactory. Some of the biological supply houses where culture dishes and agar and other ingredients of culture media can be bought are listed at the end of this chapter.

Culture media. Hundreds of different culture media are available for growing fungi, yeasts, and bacteria, and many of these media can be bought in powdered form, needing only water added to them, followed by sterilization in an autoclave or pressure cooker and they are ready to use. Most of these prepared media cost $10 or more per pound, but a pound will make from 8 to 12 liters of the final medium, enough to pour 500–600 petri dishes at the rate of 15 milliliters of agar per plate. Agar media perfectly satisfactory for growing most fungi that can be cultured can be prepared much more cheaply, and by anyone who can boil water. The ingredient that makes the medium solid when cool is agar, or agar-agar, manufactured from a certain kind of red seaweed. It is available in various forms and degrees of purity, the purer sorts being more expensive. The crude shredded agar, still redolent of the ocean and containing some diatoms and fragments of small shellfish, is much cheaper than the refined, powdered form. The formulas for several widely used culture media, and directions for preparing them, and the major uses of each, are given below.

Malt-salt agar. Weigh out 20 grams of agar — you need not be too fussy about this, since if you have 19 grams or 21 grams the fungi won't know the difference. Many people use only 15 grams, which gives a slightly less firm medium, but I prefer a good solid medium and so always use 20 grams. Put this in a pan or pot or kettle such as is used to boil vegetables in, add one liter of water (distilled water if you have it, otherwise tap water), 20 grams of powered malt extract, and 100 grams of sodium chloride, or common salt. Sodium chloride comes in various grades too, from the crude rock salt such as that used to freeze ice cream at home and to put on streets to melt the ice, to technical grade, USP (United States Pharmaceutical, which means it is pure enough to use in drugs) and CP (Certified Purity, which means that it is about as free of other minerals and so on as it is possible to get it). The rock salt costs about $3 per hundred pounds, slightly more if bought in small quantities in the grocery store, the technical grade about 10 cents a pound, and the CP grade about $3 a pound. Either the rock salt or the technical grade is satisfactory, but table salt which contains iodine, as well as some other salts to keep it free running, is not.

Boil the water, with constant stirring, until the agar is dissolved. Some of the water will have evaporated during boiling, and so you have to add water to bring the quantity up to a liter again. It is not essential that you have *exactly* a liter, a little more or less won't matter.

Pour the melted agar medium into bottles. I use 16-ounce prescription bottles for this, which with plastic screw caps cost only a few cents each. They are called prescription bottles because one of their uses is by drug stores, for liquid medicines, and if there is no firm near you that sells such bottles you can buy them from your local druggist. Divide the hot agar equally among three bottles. Put on the screw caps but do not tighten them all the way, put the bottles in an autoclave or pressure cooker, and heat them at 15 pounds pressure for 20 minutes. After the pressure has come down, tighten the caps firmly and remove the bottles. If no steam sterilizer or pressure cooker is available, this malt-salt agar need not even be pressure sterilized, since it contains too much salt for most bacteria to grow in it, and the temperature of boiling water kills ordinary yeasts and fungi. So you can stand the freshly prepared bottles of agar in hot water, heat this to boiling, and keep it boiling for 20 minutes or so, then screw down the bottle caps firmly and allow the agar to cool. However, a pressure cooker large enough to hold eight or ten 16-ounce bottles can be bought for less than $10, and since pressure sterilization is essential in the preparation of most culture media, a steam sterilizer or pressure cooker is indispensable.

It recently has been found that agar media can be sterilized by exposure to vapor of propylene oxide, so that no autoclave or pressure cooker is needed in their preparation. Freshly poured dishes of warm agar are stacked, with the covers on, on a glass plate, an open dish containing 10 milliliters of propylene oxide is set beside them, all are covered with a bell jar whose lower edge has been covered with petroleum jelly or vaseline to give a tight seal in contact with the glass plate, and left for 24 hours. The bell jar is then removed and the dishes allowed to stand 24 hours before use.

To pour the hot sterilized or subsequently remelted agar into petri dishes, put the dishes in single file or in stacks three or four dishes deep along the edge of a table. With the left hand raise the cover of a dish just far enough to insert the neck of the bottle, pour in enough agar to fill the dish about a third full, let the cover down, and proceed to the next dish. You can grasp a stack of three or four dishes conveniently with the left hand, raise the cover of the bottom one, fill it, and proceed upwards in the stack. The 300 plus milliliters of agar in one bottle will fill about 20 dishes a third full. It is just as well to use up at one time all the agar in the bottle, since if any of it is saved in the bottle it is likely to be contaminated by fungus spores that have entered the bottle with air as the agar was poured out, and will have to be discarded. If you pour unwanted agar into the sink and down the drain, run a goodly amount of water into the sink at the same time, to dilute the agar; otherwise the agar will collect and solidify in the drain trap and plug it. Also rinse out the bottle at once, so it does not sit around with a layer of agar on the inside on which fungi can grow and furnish spores to contaminate the laboratory.

If the dishes in which the agar was poured are in a room with much traffic, it pays to cover them with a bell jar or piece of clean cheesecloth so they will not become contaminated with airborne spores. If kept for a week or longer some of them will become contaminated anyway, but covering them reduces such contamination and thereby reduces waste of agar and work.

Within an hour or so the agar will be cool and fairly solid. Water will have condensed on the inside of the covers, but if the dishes are left for a day this will disappear. The agar can be used as soon as it is solid, and indeed materials can be suspended in it as soon as it has cooled somewhat and before it has solidified, but for most work it is preferable to let the dishes stand until the condensed water on the inside of the covers has disappeared.

Malt-salt agar is especially suited to the culture of storage fungi and other fungi that relish or require a high osmotic pressure. Many more colonies of some species of Aspergillus will grow from a given amount of material cultured on this medium than if they are cultured on the other media described below. The medium has been used extensively in culturing stored grains and flour, macaroni and spaghetti, condiments and spices, house dust and furniture stuffing. You can, for example, scatter over the agar surface of several dishes a small amount of flour, distributing it as evenly as possible, or shake ground black pepper onto the agar (every sample of black pepper we have tested, and we have cultured a lot of them, has been loaded with a number of species of Aspergillus) or scatter house dust or fragments of stuffing from a sofa or chair or mattress, or pieces of macaroni or spaghetti over the agar in several dishes, and leave them in the laboratory for a week or more, and see what fungi grow out. You will find that some of these common materials contain a heavy load of spores of various fungi, particularly fungi that like a high osmotic such as that provided by malt-salt agar.

Malt agar. This is very similar to the above medium, but without the salt. It contains 20 grams of agar and 20 grams of powdered malt extract per liter of medium. Preparation is as described above. Malt agar requires sterilization at about 15 pounds pressure for about 20 minutes. You will already have seen that we use the terms "agar," "agar medium," and "medium" more or less interchangeably. If you want to be pedantic you can use "agar" or "agar-agar" to designate only the seaweed product that serves to solidify the medium, and "agar medium" for the prepared or finished culture medium, but in practice few people do so, even pedants. Malt agar is somewhat acid and so is not favorable to the growth of most bacteria, but it is well suited to many fungi, and for decades it has been one of the standard media for culturing wood-rotting fungi.

Potato dextrose agar. Weigh out about 300 grams of peeled potatoes (in my early days in a plant pathology laboratory we were taught to use, always, exactly 390 grams, no more, no less, and sometimes even wore a white coat while doing so), slice these, and boil them in a liter of water until the potatoes are soft. Strain off and save the water. If you want clear agar, which sometimes is desirable, strain the water through a couple of thicknesses of cheesecloth, and wring out of the cheesecloth as much of the water as you can. Add 20 grams of agar (many people use 15 grams, others 18 grams) and 20 grams of dextrose, boil until the agar is dissolved and add water to bring it up to a liter again. This medium also requires sterilization at 15 pounds pressure for 20 minutes.

This medium, commonly designated PDA or pda (editors of scientific journals can get quite wrought-up over this PDA-pda problem, and really scour you out if you use upper case in your manuscript when they prefer lower case, and vice versa), is suitable for growing many kinds of bacteria as well as yeasts and fungi. To make acid pda, add 5 drops of 50 per cent lactic acid per 100 milliliters of the melted agar just before the agar is poured into petri dishes. If the acid is added before the agar is sterilized, or before it is remelted later, the medium will not solidify because the acid breaks down the agar gel. The acid makes the medium unfavorable for most bacteria, but many fungi grow on it very well, and it long has been one of the standard media in plant pathology laboratories throughout the world.

Czapek's agar. This medium (pronounced SHAH-pek) is a bit more complex than those described above, but the five salts it contains are likely to be present in even a modestly equipped high school chemistry laboratory. To make it, weigh out $NaNO_3$ (sodium nitrate), 3.0 grams; K_2HPO_4 (dibasic potassium phosphate), 1.0 gram; $MgSO_4 \cdot 7H_2O$ (magnesium sulphate), 0.5 gram; KCl (potassium chloride), 0.5 gram; $FeSO_4 \cdot 7H_2O$ (ferric sulphate), 0.01 gram; sucrose (any good grade of household sugar), 30 grams; agar, 20 grams. Put the agar and salts in a liter of water, boil with constant stirring until the agar is dissolved, bring up to a liter again, add the sugar and stir until it is dissolved, put in bottles, and sterilize at 15 pounds pressure for 20 minutes. For some of our work with storage fungi we use 300 grams of sucrose, instead of 30 grams, per liter.

This medium is used for the identification, by cultural characters, of many species of Aspergillus and Penicillium. That is, on this medium, a given species of Aspergillus will form colonies that are so characteristic in color and conformation that the species can be identified without much difficulty. Also on this medium, especially if it is made with 200 or 300 grams of sucrose per liter, some of the species of Aspergillus that are encountered commonly in stored grains, flour, pepper, and other food products and in house dust and furniture stuffing form very brightly colored colonies — red, green, yellow, white, brown, or black according to the species — and when a number of these colonies grow on the same petri dish they make a very good show.

Water agar. Put 20 grams of agar in a liter of water, boil until the agar is dissolved, pour in bottles and sterilize. When poured into petri dishes, fill the dishes at least half full, so they will not dry out rapidly, since you may want to keep some of the subsequent cultures for several weeks. Agar itself is relatively unnourishing, and so this medium supports only a sparse growth of fungi, which is one of its virtues. If bits of dung of horses, cows, or other herbivores are placed on this agar in petri dishes, two pieces of about a cubic centimeter each on opposite sides, and kept in a relatively cool room, a number of interesting fungi will appear in succession.

Pilobolus (discussed above at pages 80–85) usually grows out within a couple of days; its stalks, with sporangia on top, can be observed beautifully in all stages of development. Sometimes successive crops will be produced daily for a week or more. In late afternoon or early evening you can see the stalks grow out, filled with yellow protoplasm, all of them pointing at the source of brightest light. By late evening black sporangia begin to form on these stalks, and by morning the subsporangial swelling that serves as a lens to keep the sporangia pointing at the light is visible. All this is illustrated after a fashion in the drawing on page 83. If at that time you turn the dish a quarter way around, so that the light strikes the dish from a different direction, the stalks will bend sharply, just below the subsporangial swelling, and aim the tips at the light again. Frequently the mycelium of Pilobolus will grow out into the agar beneath or just beside the dung, and if you turn the dish over and put it on the stage of the microscope under low power magnification you can see protoplasm streaming rapidly from the smaller branches into the large main branch from which a stalk will grow up during the night.

Within a week small fruit bodies of Ascobolus, a miniature cup fungus, may grow out on the surface of the dung, and shoot their spores against the cover of the dish. Later the perithecia of Pleospora appear in large numbers on the dung and on or in the agar beside it; the necks or snouts of these perithecia point toward the light, and the dark spores are shot out in groups of eight. Fruit bodies of a small species of Coprinus, a gilled fungus with black spores, may appear next, a crop growing up every night to shed their spores by early morning. All of these three fungi are illustrated in the frontispiece. Very often, after the dishes have been sitting for a week or 10 days, nematodes swarm out from the dung and over the surface of the agar, and these may be followed by nematode-trapping fungi. With a microscope the actual trapping followed by the death struggles of the nematodes and their digestion from the inside out by the fungus can be observed.

If the dung of frogs, toads, or snakes is placed on water agar, stalks of Basidiobolus grow out within 24 hours, each stalk surmounted by a single large spore that grows to its full size within an hour and is then shot off toward the light. The spores that land on the agar will germinate and send up another stalk, the tip of which swells up, points at the light, forms another spore, and shoots it away. Basidiobolus and most of the nematode-trapping fungi can be cultured easily on acid pda or malt agar. Pilobolus, Ascobolus, and Pleospora can be cultured too if you make dung agar (water agar with a small amount of animal dung added to it before sterilizing), put on this the spores of the fungus you want to cultivate, and keep the culture dishes at a temperature of 40–42 degrees Centigrade (105–110 degrees Fahrenheit) for 48 hours. The spores of these fungi are adapted to passage through the intestines of animals, where they are exposed to a temperature of about 40 degrees Centigrade for a day or two, and in their way they are specialists with many nice adjustments to their environment. Once the spores have been stimulated to germinate, by exposure to high temperature, the cultures can be kept at room temperature.

Culturing materials on agar media. Some of the materials you can culture have been suggested above, but there are many more. One interesting project is to determine the number and kinds of culturable fungus spores in the air within buildings and outdoors. To do this, remove the cover of a petri dish that contains agar, leave the agar exposed to the air for 15 minutes, replace the cover, and write the exposure time and place on the dish cover with a wax pencil, or number the dish and record in your notes the number and the time and place of exposure. Leave the dish in the laboratory until the fungi grow out, then count and record the number of colonies of each kind.

To find out what effect the composition of the culture medium has on the number of colonies that develop from spores that fall onto the agar, expose one or preferably two dishes of each kind of medium — malt-salt, malt, potato dextrose, and Czapek's agar in each place and at the same time.

If you expose sets of dishes in this way in your home at different times and in different rooms, you can find out whether the spore load is heavier at some times or places than others. If you expose some on the floor beside a sofa or overstuffed chair, then sit on the sofa or chair several times and so expel air and fungus spores from the stuffing, you may get an exceedingly large number of colonies. In this way you may have demonstrated why some of the people who occupy or visit your home get sniffles and sneezes and asthmatic symptoms of varying severity whenever they sit on such furniture. The allergists who work with respiratory allergies caused by airborne spores of fungi seldom encounter the species of Aspergillus that in our experience are likely to be so abundant in furniture stuffing and in other things around the average home. In the first place these men normally expose their culture dishes outside the windows of their own offices or laboratories, and so sample the culturable spores in the outside air at that time and place. The number and kind of fungus spores within a given home or place of work may be quite different from the number and kind in the outside air; the source of spores that give respiratory troubles to many people is *within* the home or place of work. Second, few of those who have sampled air for fungus spores have used a medium such as malt-salt agar that will disclose the presence of spores of *Aspergillus restrictus* and some of its near relatives, fungi that are much more abundant within some buildings, including homes, than they are out-of-doors. There has, indeed, been very little work of this sort done in homes, and so with a well-planned and conscientiously carried out study you might come onto something entirely new, especially if you keep your eyes and mind open.

Cultures of wood-rotting fungi and mushrooms. Many wood-rotting fungi will grow readily in culture on malt or acid potato dextrose agar. To make a culture from decayed wood, split off a piece of the wood near the edge of the advancing decay, where the obviously decayed and the evidently sound wood meet, and take this to the laboratory. The piece should be of a size convenient to work with, say a couple of inches wide, half an inch or so thick, and six inches to a foot long. Flame the outside of this piece with an alcohol lamp or Bunsen burner, just heavily enough to char it slightly. Then cut off a piece of the inner surface to expose the decayed wood anew. Pass a razor blade or knife blade through the flame of an alcohol lamp or Bunsen burner, just long enough to sterilize the blade, make several cuts about ⅛ inch apart with the grain of the wood, and several more cuts the same distance apart across these. Pass the points of a pair of tweezers through the flame to sterilize them, pry up half a dozen small pieces of wood, and put these on malt agar in a petri dish. Repeat this, so you have two dishes, each with half a dozen pieces of decayed wood on the agar. Leave these under a bell jar or other covering and examine them occasionally. Most wood-rotting fungi grow more slowly on agar than do many of the common molds, and form colonies of white or brown mycelium without masses of powdery spores. So if the colonies that grow from the wood are green or yellow or black and have masses of powdery spores it is likely that you do not have the fungus responsible for the rot. Fungi that rot wood frequently are accompanied or followed in the wood by bacteria and by various other fungi. To make cultures from the fruit bodies or conks, the spore-producing structures of the fungi, select fresh, sound fruit bodies, split or cut them open in the laboratory, and with flamed tweezers take out bits of the interior mycelium and put these on malt agar in petri dishes. Usually colonies will grow out within a few days. Many wood-rotting fungi can be identified by means of their cultural characters, but this is a somewhat advanced and specialized business.

Some but by no means all mushrooms or gilled fungi can be cultured from fruit bodies in the same way. Pick fresh, sound mushrooms, cut off the stems so that no dirt is carried into the laboratory, split a cap down through the center and with sterile tweezers take out small bits of mycelium just above the gills or just where the flesh of the cap joins the stem, and put these on malt or acid potato dextrose agar in petri dishes. Mycelium of some of the mushrooms grows out from such pieces in a day and some of these cultures, if left long enough, will form mushrooms similar to but smaller than the originals from which the cultures were made. There has been a good deal of rather basic work done on genetics of fungi with some of the mushrooms cultured in this way, especially with those that grow on wood and with small species of Coprinus that can be found after every rain on any manure pile.

Cultures of plant parts infected with fungi. Many of the fungi that cause plant diseases can be cultured, including all those that cause damping off and seedling blights, described above, those that cause rots of fruits and vegetables, and those that inhabit seeds.

For most of this work you need to have some sort of liquid disinfectant in which to wash the plant or plant part you want to culture, to eliminate contaminants from the surface of it. A household or laundry bleach of which the active agent is sodium hypochlorite is satisfactory for this. Usually this comes as a 5.35 per cent solution of sodium hypochlorite; to make an effective surface disinfectant for laboratory use, dilute one part of this with four parts of water.

Cut decayed stems and roots of plants into sections about ¼ inch long, put a dozen such sections in a petri dish, add enough of the disinfectant to cover them, shake the dish for about a minute, pour off the liquid, and flush the pieces with sterile water, which has been heated at 15 pounds steam pressure for 20 minutes, and subsequently cooled, in bottles like those used for the culture media. With flamed tweezers put half a dozen pieces on each of two plates of malt agar or acid pda and leave the dishes in the laboratory until the fungi grow out.

Seeds of corn, barley, wheat, rice, or other agricultural plants can be surface disinfected and cultured in the same way. Sometimes it is desirable to culture one lot of seeds on malt agar or acid pda to find out what field fungi are in them, and another lot on malt-salt agar to detect storage fungi.

In all this culture work with agar media, cleanliness is essential — freshly washed hands; a freshly washed smock or a rubber apron that can be sponged off before the work is undertaken; the desk or table where the work is done and the dishes kept washed with hot soapy water; windows and doors kept closed during the time cultures are made, to avoid drafts that carry airborne spores; the floor as clean as possible.

A Fungarium

A fungarium can easily be made that will serve the same purpose for fungi that an aquarium does for water plants and fish, or a terrarium does for land plants and animals. With an imaginative approach you can make a most attractive display of many sorts of fungi, and if old specimens are removed when their usefulness is past and replaced with fresh ones, a number of living specimens can be kept on view more or less all the time. Don't throw out everything that is beginning to get moldy or overgrown by fungi, because after all fungi are what you are trying to grow.

I have used for this an old discarded aquarium that no longer would hold water, but a perfectly suitable fungarium can be made with little cost. It should be about 16 inches wide, 24 inches long, and 14 inches high. The bottom can be of wood or metal, and need not be watertight. Corner posts of wood should be grooved vertically to hold glass sheets cut to make the sides and ends. The top should be covered with a sheet of glass just a bit larger than the top dimensions so that it can be lifted off and put on easily.

Put a layer about four inches deep of sawdust or a mixture of soil and sand or a mixture of soil, sand, and partly decomposed compost in the bottom. Add water enough to make this moist but not sopping wet, and set the box in a cool room, out of direct sunlight.

To get quantities of Pilobolus, the cap thrower described in Chapter 4 and also earlier in this chapter, collect a pint of fresh horse dung, or dung of almost any herbivore (we have consistently got excellent crops of Pilobolus, Ascobolus, and Coprinus from buffalo dung; if your city has a zoo, you might try the dung of elephants, giraffes, and other outlandish animals, just to see what you get). Put the dung on the bottom in one corner of the fungarium, preferably farthest from the light. Ordinarily Pilobolus will appear in quantity within three days and shoot its sporangia off toward the source of brightest morning light. Successive crops may be produced daily for a week or more, and the discharged sporangia will form a dark patch on the inner side of the glass wall. A number of interesting fungi will follow Pilobolus, especially Coprinus, a genus of mushrooms with black spores, which may produce successive crops every morning for weeks.

Young mushrooms growing on wood or on the ground can be transferred to the fungarium with a moderate-sized hunk of soil or wood on which they are growing, and may continue to develop for some time. A half dozen sorts of these of different sizes, shapes, and colors make an attractive show. In the quiet air of the fungarium it sometimes is possible to see clouds of spores floating away from these fruit bodies. Some of the larger cup fungi that are common almost everywhere in spring will remain fresh for some time in the fungarium, and will discharge clouds of spores intermittently, especially when they are blown upon or touched. Lichens will remain fresh for weeks.

With two such fungariums, or fungaria, one can be used for fungi such as those described above, the other for fungi that grow on and spot or decay or discolor various sorts of materials. Into this second fungarium, also on a base of moist sawdust, soil, or compost, put such things as paper in the form of books or bulletins, a few different kinds of fabrics and pieces of rope, a couple of slices of bread, some whole fruits or vegetables, an old shoe or leather glove, pieces of bark two or three inches wide and eight inches long from a number of different kinds of recently dead trees, some pieces of painted wood and insulating board. Keep the fungarium covered and see what fungi develop on the different materials.

Sources of Material for Making Cultures

Agar and other ingredients of culture media: Difco Laboratories, Inc., Detroit 1, Michigan. Ben Venue Laboratories, Inc., Bedford, Ohio. Baltimore Biological Laboratory, Inc., Baltimore 18, Maryland.

Petri dishes and other apparatus: Geo. T. Walker and Co., Inc., 2218 University Ave. S.E., Minneapolis 14, Minnesota. Chicago Apparatus Co., 1735 North Ashland, Chicago 22, Illinois. Fisher Scientific Co., 717 Forbes, Pittsburgh 19, Pennsylvania; 635 Greenwich, New York 14, N.Y.; 7722 Woodbury, Silver Spring, Maryland; 2850 S. Jefferson, St. Louis 18, Missouri; 904 St. James, Montreal, Canada; 245 Carlaw, Toronto 8, Canada. E. H. Sargent & Co., 4647 West Foster, Chicago 30, Illinois; 1959 East Jefferson, Detroit 7, Michigan; 5915 Peeler, Dallas 9, Texas. Central Scientific Co., 1700 Irving Park Road, Chicago 13, Illinois; 237 Sheffield, Mountainside, New Jersey; 79 Amherst, Cambridge 42, Massachusetts; 3232 Eleventh N., Birmingham 4, Alabama; 6446 Telegraph Road, Los Angeles 22, California; 1040 Martin, Santa Clara, California; 7275 St. Urbain, Montreal 14, Quebec, Canada;

146 Kendal, Toronto 4, Ontario, Canada; 1206 Homer, Vancouver 3, B.C., Canada; 130 Sparks, Ottawa, Ontario, Canada.

Additional Reading

H. L. Barnett, *Illustrated Genera of Imperfect Fungi*, 2d ed. Minneapolis: Burgess Pub. Co., 1960. 225 pp. Keys, descriptions, and illustrations of 462 genera of Fungi Imperfecti.

C. M. Christensen and Mary Ann Swaebley, "Identification of Cultures of Common Airborne Fungi." Mimeographed. 1951. 8 pp. Available without charge from the Department of Plant Pathology, University of Minnesota, St. Paul 1, Minn.

Leander F. Johnson, *et al.*, *Methods for Studying Soil Microflora — Plant Disease Relationships*. Minneapolis: Burgess Pub. Co., 1959. 178 pp.

A. J. and R. S. Riker, *Introduction to Research on Plant Diseases*. St. Louis and Chicago: J. S. Swift Co., 1936. 117 pp.

George Smith, *An Introduction to Industrial Mycology*, 4th ed. London: Edward Arnold Publishers, Ltd., 1954. 378 pp. Keys, descriptions, and illustrations of fungi; laboratory technics; maintenance of cultures of fungi; control of fungi; industrial uses of fungi; mycological literature.

United States Department of Agriculture, publications on market diseases of fruits and vegetables: *Potatoes*. Miscellaneous Publication 98. 1936. *Tomatoes, Peppers, Eggplants*. Miscellaneous Publication 121. 1932. *Peaches, Plums, Cherries and Other Stone Fruits*. Miscellaneous Publication 228. 1937. *Apples, Pears, Quinces*. Miscellaneous Publication 168. 1933. *Crucifers and Cucurbits*. Miscellaneous Publication 292. 1938. *Grapes and Other Small Fruits*. Miscellaneous Publication 340. 1939. All these include excellent colored plates of the diseases and rots they describe. They can be obtained from the Superintendent of Documents, Government Printing Office, Washington 25, D.C., for from 20 cents to 45 cents each.

Summary of the Classification of the Fungi

THAT part of mycology dealing with the classification, or taxonomy, of the fungi has several different tasks. The most important of these are (1) to describe the individual kinds or species of fungi, briefly, clearly, and distinctly; (2) to name these species; (3) to arrange the described and named species into an orderly system. Each of these will be discussed very briefly.

The description of species. This is not always so simple as it seems. Some individual species are so distinct, so obviously different from others, that they can be recognized easily and without question as discrete and separate entities. Such distinct species make up only a very small minority of the total number. Most species of fungi grade into their near relatives by such small and insensible steps that no one can say with any degree of certainty where one species ends and another begins. The characters supposedly typical of a given species may be altered by environment, so that the same fungus may differ from place to place and from time to time. Genetic variations also are common — a given fungus may mutate, and many fungi hybridize to produce offspring which not only fluctuate between the two parental types, but may even transcend the parents in one or more characters.

Considering the amount and kind of variation proved to occur in various fungi that have been investigated thoroughly, one might think it desirable, before describing a new species of fungus, to study these known sorts of variation in some detail. Theoretically, a careful investigator, before foisting a new species upon a field already overburdened with poorly defined species, should collect the fungus from a number of different places over a period of at least several years. He should grow various collections of it either in culture or upon its natural substrate or host, under a number of different environmental conditions. He should also determine whether it hybridizes, and, if it does, the range of variation among the progeny. He also should investigate several of the closely related species with equal thoroughness. In this way he could eventually accumulate enough information to determine whether this supposedly new species was actually sufficiently different to deserve description as new or whether it might be incorporated into a species already described. Such technics have been used by some of the modern taxonomists of fungi, often with rather startling results. This experimental approach to taxonomy requires a study of the biology of the fungus involved. Many taxonomists of fungi still describe new species, and rearrange old ones, solely from a study of dead material, or from the study of one or a few collections of living material. They add more to the number of species than to our knowledge of fungi.

The naming of species. Once a new fungus is described, it must be given a Latin name. Theoretically, it retains this name for ever and always. In a beginning course in taxonomy we are taught that a scientific name of a plant or animal, even if it be one of seventeen syllables and practically unpronounceable, has at least the virtue of constancy. As we progress, we learn to our sorrow that this is not so. Many common fungi have been given at least a dozen different Latin names. Some of the common fungi have been given more than twenty different scientific names.

In some cases this has been done innocently. Different men, working in different places, have given different names to the same fungus. In most cases, though, the series of new names have been applied with full knowledge of the work that has gone before. Often the person who has to teach mycology to beginning students wonders if these taxonomists always distinguish clearly between progress and confusion. In most cases where many different names have been applied to the same fungus, very little has been added to our knowledge of the biology of the organism concerned — only the new name has been added.

Often a new name is applied to an old fungus in the attempt to make a more reasonable or supposedly more logical arrangement of that particular group, or to make the name conform to a particular author's concept of how the group ought to be organized, or to make the name conform to the Rules of Nomenclature. Only very seldom are such changes made with the idea of making it easier for students, teachers, or research workers to recognize and identify the fungus in question.

The Rules of Nomenclature, mentioned above, loom large indeed in the classification of fungi. The rules of procedure in the United States Senate are as the play of prattling children compared with the Rules of Nomenclature of the fungi. These rules have a certain deceptive simplicity that embodies a few obvious loopholes, some hidden obscurities, and a number of ambiguities — all of which add up to a tangle of Gordian knots. Why not simply cut the knots? Anyone suggesting so simple a solution obviously does not know taxonomists. So there has developed a special branch of taxonomy known as nomenclature, and some mycologists devote much of their time and talent solely to nomenclature — to the naming of fungi and to determining the legality of the names that have been applied to fungi. After all, it is essential that when we speak or write of a given fungus, those who hear or read shall know of what fungus we speak. Those who describe and name species, and those who work with the devious problems of nomenclature, contribute vitally to mycology. Both should recognize that fungi are of importance to us as living, growing organisms, not only as dead and dried specimens in a herbarium. Some of them do.

The arrangement of taxonomic systems. One difficulty inherent in arranging the fungi into an orderly system is that so many of the fungi are so simple in structure. It is a truism in taxonomy, if seldom expressed, that the simpler the organism, the more complex and difficult its classification. This is as true of, say, different kinds of tools or different kinds of fabrics as it is of fungi. There are, in the arrangement of fungi, two different and somewhat conflicting bases, or modes, of procedure. One is to group together those fungi that look most alike, to facilitate recognition and identification. The other is to arrange the system according to the supposed evolution or phylogenetic development of the fungi, so that those fungi most nearly related may be near one another in the system of classification.

Most of the classifications of fungi so far produced have been combinations and compromises of the two. It is not easy to arrange the fungi according to their phylogenetic development — their course of evolution — when we know almost nothing about how fungi have evolved. The fungi are so small and so perishable that their remains have not been preserved in the stony deposits that have given us so much of the evolutionary background of our prehistoric plants and animals. We simply do not know what the fungi have evolved from; or, in many cases, which of the present forms are most closely related; or, in a given group, which ones are highest and which ones lowest, from an evolutionary standpoint. Many fungi are out on the end of different evolutionary limbs whose connection with the parent tree is lost in the mists of antiquity. Those who classify them are out on the same limbs. While they may recognize this, they seldom state it explicitly. Different authorities have different ideas as to the arrangement of species within genera, genera within families, families within orders, orders within subclasses, subclasses within classes, and even as to what classes or organisms should be included in the fungi.

It is almost impossible to find two systems devised by different authorities that agree. It is not unusual to find one authority giving two different systems at different times in his life. The beginning student often is confused and bewildered to find a given fungus listed in one family and order in one text, in a different family of the same order in a second text, in the same family but in a different order in the third text, and in a different family and different order in the fourth text. He must recognize that this merely expresses the confusion and bewilderment of those mycologists who write texts. Those who do not write texts are no less confused and bewildered, but relatively few of them devote much time and thought to the construction of imaginary courses of development of real or imaginary fungi. Given this confusion, it perhaps is unfortunate that the taxonomists of fungi cannot agree on a single general arrangement. Since we have no evidence one way or the other as to how fungi have evolved, it would do no harm scientifically, would commit no scientific violence, merely to agree on one system. An arbitrary system to be sure. All systems so far devised have been largely arbitrary. It would make it so much simpler, easier, and more convenient if all those who teach mycology, anywhere in the world, could have at least the same general framework to hang the fungi on. Nor would this stultify the discipline of mycology. Progress in this field will not come from devising new and different frameworks on which to hang the fungi, or from a new design of pigeonholes in which to arrange dead and dried specimens of fungi, but from a study of the biological processes of living fungi.

The classification given below follows essentially that given in G. W. Martin's *Families of the Fungi*. This one is chosen partly because it seems a relatively simple, reasonable, teachable, and learnable system, partly because Dr. Martin carries considerable weight in the mycological world, partly because it has been accepted by Ainsworth and Bisby in their *Dictionary of the Fungi*. This *Dictionary of the Fungi* is published by the Imperial Mycological Institute, in England, one of the world's foremost mycological institutes. It is used by students of fungi throughout the world. If Dr. Martin's classification is not accepted by all, it certainly should be more familiar to more mycologists, and be more widely followed, than some of the other arrangements that have been devised by this or that author and followed by no one else at all.

With this introduction, we proceed to the classification of the fungi.

The Plant Kingdom is divided into four large groups, called Divisions, one of which is the Thallphyta, or Thallophytes, or thallus plants. This Division is in turn divided into several Subdivisions, one of which is the Eumycetes, or true Fungi. As stated in Chapter 1, nearly all of the fungi, or Eumycetes, have in common the three simple characters of: (1) no chlorophyll, (2) a growing or vegetative portion made up of mycelium, and (3) reproduction by means of spores.

The Subdivision known as Eumycetes, or Fungi, is further divided into four Classes, each of which will be described briefly.

Class 1. Phycomycetes. The name Phycomycetes means *algal fungi*. They were given this name because, when they first were studied intensively, between 1850 and 1900, a few of them were found to resemble certain green algae. A theory that was popular for longer than it should have been postulated that they had evolved from the green algae. No one really knows where, how, when, or from what the Phycomycetes evolved, nor are we likely to find out. They include about 1,000 species, in 235 genera. About one half of the described genera include only a single species.

The Phycomycetes are supposed to be characterized by: (1) Mycelium without crosswalls — nonseptate — so that the entire thallus, or growing plant, consists of a single cell. In general this is so, but there are some exceptions. Structures specialized for spore production often are delimited by crosswalls, and in the old mycelium of some species crosswalls may be present at irregular intervals. (2) Asexual spores produced in sporangia. Sporangium means *spore case*. In most of the Phycomycetes the terminal portion of a special hypha swells, the nuclei in it increase in number, and a large number of one-celled spores are formed — just about as quickly as that. A spore can germinate, produce a tangle of mycelium of visible size, and then produce a crop of spores, all within less than forty-eight hours. In the water-inhabiting Phycomycetes these sporangiospores are ciliated zoospores, or swimming spores. In the terrestrial species, with aerial sporangia, the spores are mostly single oval or spherical cells adapted to dissemination by air. The sporangia of Phycomycetes never are aggregated into complex fruit bodies, as often is the case with asexual spore-producing hyphae of the higher fungi. The Phycomycetes usually are divided into two Subclasses: (1) Oomycetes, in which the sexual spores are oospores, formed by the fusion of a small antheridium with a large oogonium. After fertilization, one to many oospores form within the oogonium. (2) Zygomycetes, in which the sexual spores, zygospores, are formed by the fusion of gametangia of similar size and shape. Each fusion results in only a single zygospore. The sexual spores are not encountered commonly in nature, and in practice we recognize most Phycomycetes from the asexually formed sporangia and sporangiospores. The Phycomycetes mostly inhabit water, moist soil, and succulent parts of plants.

Class 2. Ascomycetes. All Ascomycetes produce ascospores, and any fungus that produces ascospores is an Ascomycete. With the exception of the yeasts and some of their relatives, most Ascomycetes have a well developed and regularly septate mycelium, but an Ascomycete cannot be recognized by its mycelium alone. Most of the Ascomycetes produce asexual spores, called conidia, either in definite fruit bodies or on special branches scattered on the mycelium.

Sexual reproduction is by means of ascospores in asci. A typical ascus is a long, cylindrical or club-shaped cell that, at maturity, contains eight ascospores. The ascospores vary in shape from long and threadlike to oval or spherical, and in cell number from one to many. Many of the Ascomycetes forcibly expel the ascospores, sometimes in visible clouds. This, of course, facilitates aerial dissemination of the spores. Many of those Ascomycetes that are associated with insects and depend upon the insects for the dissemination of their spores do not shoot their spores out into the air. These usually produce their ascospores within a closed fruit body, with a pore at the top. At maturity the walls of the asci decompose or dissolve into a sticky, thick liquid. When this absorbs water it swells, causing the spores to be exuded in a gelatinous matrix that adheres readily to insects and other animals that come in contact with it.

In most of the Ascomycetes the asci and associated cells are aggregated into structures called fruit bodies. These vary in size from microscopic to several inches in height or diameter. About 40,000 species of Ascomycetes have been described, but Ainsworth and Bisby, mentioned above, consider only 12,000 of these to be legitimate species. Ascomycetes are not common in lakes, rivers, or oceans, but they are exceedingly abundant in a great variety of land environments.

The Ascomycetes are usually divided into two Subclasses: (1) Hemiascomycetes (half Ascomycetes), whose asci are borne singly or scattered on the mycelium, not aggregated into fruit bodies, and (2) Euascomyces (true Ascomycetes), all of which have their asci aggregated into fruit bodies. Some of those Ascomycetes that produce conidia regularly and ascospores rarely are classified with the Fungi Imperfecti, or in both the Ascomycetes and the Fungi Imperfecti. While at first glance this may seem rather confusing, it merely expresses the uncertainty and irregularity that occur in the fungi themselves; and when one gets accustomed to the fact that a given fungus may be placed in either one or both of two different groups, it is no longer troublesome.

Class 3. Basidiomycetes. All fungi in the Class Basidiomycetes bear basidiospores on basidia, usually four spores per basidium. In most cases these basidia are borne on or within fruit bodies large enough to be visible to the naked eye, and many of the Basidiomycetes can be distinguished easily without the use of a microscope.

The mycelium of all Basidiomycetes is regularly septate; the mycelium of many species has a peculiar "clamp connection" that grows out of the cell just ahead of a wall, and back down into the cell just back of the wall. Any fungus with such clamp connections on its mycelium is definitely and positively a Basidiomycete, but not all members of the Class have them. The sexual spores are basidiospores, always borne on small stalks on the outside of the basidium, usually four spores per basidium, as stated above. Nearly all basidiospores consist of just a single cell, and almost all of them are forcibly abjected from the stalk on which they are borne, a process that involves some delicate mechanical and physiological adjustments. Many Basidiomycetes bear asexual spores, or conidia, of various kinds, either scattered on the mycelium or in or on special fruit bodies. Like the Ascomycetes, they are seldom encountered in aqueous environments, but are common nearly everywhere else.

The Basidiomycetes are divided into two Subclasses: (1) Hemibasidiomycetes, or Heterobasidiomycetes, characterized by basidia with longitudinal or transverse walls that divide each basidium into four cells, each of which produces a single basidiospore. (2) Homobasidiomycetes, characterized by one-celled basidia, each of which bears its spores on short stalks at the tip. According to Ainsworth and Bisby's *Dictionary of the Fungi*, the Basidiomycetes include some 13,500 species.

Class 4. Fungi Imperfecti. The fungi in this large group produce only conidia, or asexual spores. At the time when fungi were first being investigated thoroughly as living, growing organisms about a hundred years ago, some of the more obvious and common Ascomycetes and Basidiomycetes were studied in detail. They were found to produce both asexual spores, or conidia, and sexual spores, either ascospores or basidiospores. It gradually came to be assumed that *every* fungus had, or should have, sexual reproduction somewhere or other in its life cycle, that under certain conditions a fungus which ordinarily reproduced only by asexual spores would produce sexual spores. Those fungi that did not produce sexual spores, but reproduced only by conidia, were considered to be "imperfect," and were thrown together into the Fungi Imperfecti. It was assumed at the time that this group would be only a temporary one, for convenience only, and that gradually most of these imperfect fungi would be placed in natural groups of Ascomycetes or Basidiomycetes. Those Phycomycetes which produce only asexual spores — sporangiospores — are known to be Phycomycetes both from their nonseptate mycelium and their peculiar type of asexual spores, and so presumably none of the Phycomycetes have been gathered into this heterogeneous group of Fungi Imperfecti. As time went on, it became apparent that many of these imperfect fungi never produced sexual spores. Either they never had enjoyed the supposed biological benefits of sex, or else had lost them. In any case, most of them produce only asexual spores now. They include about 10,500 species, according to Ainsworth and Bisby, although about 30,000 supposed species have been described, or at least 30,000 different names of species have been published in mycological literature. They are classified into orders and families according to the way the conidia are produced, and into genera and species according to the size, shape, color, and other characters of the spores. They are found literally everywhere and at all times.

In addition to the above four groups of fungi, there is another and smaller one, known as Mycelia Sterilia, fungi which produce no spores whatever. Only a few genera are included, but some of these are very widespread geographically and common in occurrence. In spite of their lack of formal reproductive parts, they seem capable of successful competition.

The above has given in only very brief form the major characteristics of the four Classes of fungi and of the Subclasses within each of those that boasts Subclasses. We shall now summarize the characteristics of the principal orders within each of the Classes and Subclasses.

Class Phycomycetes

SUBCLASS OOMYCETES

Order 1. Chytridiales. The mycelium of typical Chytridiales consists of a tuft of a few branches that taper to invisibly fine points. Usually each tuft of mycelium is a unit unto itself, and it does not give rise to secondary colonies, as does the mycelium of most other fungi. A few Chytrids have been described in which the first clusters of mycelium give rise to others, and many more of them may exist than are yet known. A single cluster of mycelium usually produces a single sporangium on the surface of the host or substrate, and at maturity zoospores are liberated from this.

The typical life history is simple. A zoospore germinates to form a single hypha that invades the host or substrate on which the plant lives. If it invades a living plant or animal, the mycelium usually is restricted to a single host cell. Eventually this mycelium produces a sporangium on the surface of whatever the Chytrid is growing on, a number of zoospores develop in this sporangium, escape through a terminal pore, swim around for a time, and repeat the process. It seems a simple and rather uninteresting life, and it probably is.

But Chytrids do get around. Most of them live in water or in moist soil. Many are parasitic on other fungi, including other Chytrids, on such simple animals as protozoa and rotifers, on algae, higher plants, and on various animals. A few are important parasites of crop plants. Some genera are found from the tropics to the arctic and antarctic regions. Some are almost universal inhabitants of moist soil. One can isolate some of these relatively unknown fungi readily from almost any bit of soil by placing a small amount of it in water and "baiting" the water with hair, cellophane, chitin, or meat. Some of them have a very restricted diet, and may grow only on chitin. They apparently are common in the soil, and may have a more important part in the economy of the soil than we now know. The genus Synchytrium, often included in the Chytridiales (and often in the Plasmodiophorales, in the Slime Molds), includes about 75 species, most of them parasitic on various wild and cultivated seed plants, on which they cause galls or overgrowths. *Synchytrium endobioticum* causes a destructive disease of potatoes, commonly known as black wort. *Physoderma zea-maydis* is a common parasite of the leaves of maize in the southeastern United States, and *Urophlyctis alfalfae* causes overgrowths on the roots of alfalfa in the western United States.

Order 2. Saprolegniales. The fungi in this order have well-developed mycelium, sometimes of such large diameter that it can be seen easily with the naked eye. The asexual spores are produced in long, club-shaped sporangia. Usually these asexual spores are zoospores, which escape from the sporangium through a terminal pore, dozens of them sometimes pouring out in a few seconds, like students out the lecture room door when the lecture is over. These zoospores may go through several stages of development before they finally germinate to form mycelium. In sexual reproduction large oogonia are formed, are fertilized by one or several antheridia, and from one to twenty or more oospores develop within a single oogonium. Sometimes the oospores, the sexual spores, develop without any fertilization. It has been proved that in some species a number of different sex hormones may be involved in the development of oogonia and antheridia and in subsequent fertilization, so the process is by no means so simple as it appears. Also, a given strain, derived from a single spore, may behave as a female when mated with one "opposite" strain, and as a male when mated with another "opposite" strain. This fluctuation from one sex to another is not uncommon among the fungi, and no moral opprobrium is attached to it. Biologically it is very useful. In the life cycle of a typical member of the Saprolegniales, a zoospore germinates to form mycelium. This grows and produces a large number of sporangia and sporangiospores in a short time, in this way increasing and multiplying rapidly. Oospores are formed within a few days, and these are capable of surviving low temperature or other unfavorable conditions—in fact, they often will not germinate until they have matured for weeks and have been subjected to low temperature. They probably serve to carry the plant through the winter in a dormant state.

The Saprolegniales grow chiefly in water, and some of them can be isolated readily from the water in ponds, lakes, rivers, and even the water in birdbaths simply by "baiting" a small quantity of such water with boiled hemp seeds, berries, dead flies, or bits of meat. Some of the Saprolegniales are common in the soil, where they grow saprophytically on plant remains and occasionally as parasites in plant roots. Most of the common members encountered are in the family Saprolegniaceae, which includes such ubiquitous genera as Saprolegnia, Achlya, and Aphanomyces.

Order 3. Peronosporales. The fungi in this order have fine, abundant mycelium and spherical or oval sporangia. The obligately parasitic members of the Peronosporales bear large numbers of sporangia on tufts of upright, branched stalks, the sporangiophores. In most of the genera the sporangia are readily detached from their stalks and disseminated by the wind. The sporangia germinate either by the formation of one or more zoospores — swimming spores — that soon come to rest and form a germ tube that quickly develops into mycelium; or the sporangia themselves form a germ tube. A one-celled sporangium that normally is disseminated by the wind and germinates by a germ tube is scarcely distinguishable from a conidium, and some of the sporangia produced by fungi in this order are regularly referred to as conidia.

The order includes 3 families. The Pythiaceae have sporangiophores that are very similar to the vegetative mycelium, and most of them are saprophytes or facultative parasites in the soil or on the roots or other succulent parts of plants. The principal genera in this family are Pythium and Phytophthora. Both genera include economically important plant parasites of almost worldwide distribution. Some species of Pythium attack and decay the roots of more than one hundred different species of higher plants. *Phytophthora infestans* causes late blight of potatoes.

The family Peronosporaceae includes the downy mildews, obligate parasites of higher plants. They produce large numbers of sporangia on the branched tips of sporangiophores. Usually successive crops of these are produced at night, growing out through the stomata of infected host plants at night. The sporangia of most Peronosporaceae germinate to form zoospores, which soon form germ tubes. In one genus, Peronoplasmopora, the sporangia germinate either by zoospores or germ tubes. In Peronospora they germinate only by germ tubes. Oospores are formed by most species, within the infected host tissue. These fungi overwinter either by dormant mycelium or by oospores, or both.

The family Albuginaceae includes only a single genus, Albugo, or Cystopus, made up of about 25 species. All of these are obligate parasites of higher plants, and are known as white rusts. They produce chains of sporangia, in pustules or sori, just beneath the epidermis of the infected host. When mature, these sporangia are disseminated by the wind. *Albugo candida* parasitizes some 40 species of crucifers such as radish, mustard, cabbage, and water cress. In the fall, numerous oospores with characteristically warted walls are formed in the host tissue and persist over winter. In the spring each oospore germinates by forming a vesicle in which zoospores develop. When liberated, these probably are carried to the young leaves of the host plants by splattering raindrops.

SUBCLASS ZYGOMYCETES

Theoretically the fungi in this Subclass are distinguished by the production of zygospores. Actually zygospores seldom are seen either in nature or in ordinary cultures, and most of the members are recognized by the kind of sporangia they bear. Two orders are included.

Order 4. Mucorales. A fungus with black sporangia borne on long stalks visible to the naked eye is almost certain to be a member of the Mucorales. There are, however, many variations of the standard pattern of spore production, and there is some question whether certain fungi now included in the Mucorales actually are Phycomycetes or might better be included in the Fungi Imperfecti. The order includes 5 families, distinguished from one another on the basis of the structure of the sporangiophores and sporangia. The most common members probably are Mucor and Rhizopus, in the family Mucoraceae, and Pilobolus, in the family Pilobolaceae.

Order 5. Entomophthorales. Most of the fungi in this order are parasitic upon insects. The mycelium grows within the bodies of the hosts as large individual cells or clusters of cells, called hyphal fragments. Asexual reproduction is by means of single, terminal conidia that are forcibly shot off (with the exception of Massospora). Zygospores may be formed within the infected host. Three families are included. The best-known members of the group are Empusa, parasitic on a variety of insects, and Basidiobolus, which grows in the intestines of frogs, toads, lizards, and snakes and discharges its spores from the freshly deposited excrement of these animals.

Class Ascomycetes

SUBCLASS HEMIASCOMYCETES

In this group the asci are borne scattered on the mycelium, never aggregated into fruit bodies. Two orders are included.

Order 6. Endomycetales. The distinguishing character of this order is that the asci are formed directly from the fusion of two vegetative cells. Three families are included: the Ascoidiaceae, with sparse mycelium and many-spored asci; the Endomycetaceae, with asci on well-developed mycelium; and the Saccharomyceta-ceae, which comprises all of the typical yeasts. Yeasts grow by means of budding, although a few of them have some mycelium also. Yeasts are common and nearly ever-present in soil, decaying vegetation, on living plant parts such as flowers, fruits, and seeds, and on and in some animals. *Saccharomyces cerevisiae,* the common baking and brewing yeast, is one of the important economic plants of the Western Hemisphere. *Torula utilis* has been grown to some extent as a source of food and feed, particularly in those regions where proteins are in chronically short supply, but it is not likely to become a major source of human food in the forsee-able future.

Order 7. Taphrinales. As they occur in nature, all the fungi in this order are obligate parasites, but most of them can be grown in the laboratory on special media. They produce asci in a layer on the surface of the infected host. None of them are known to produce conidia. The growth of the mycelium within the hosts induces the formation of galls, overgrowths, and distortions of various kinds, and some of them are responsible for considerable economic loss.

SUBCLASS EUASCOMYCETES

All of the fungi in this group produce their asci in fruit bodies of one sort or another. The two general types of fruit body are an apothecium, which is typically cup-shaped, or saucer-shaped, with the asci in a layer on the upper side, and a perithecium, a more or less spherical structure with the asci arising from the inner wall and the spores escaping through a pore at the top. Fourteen orders are included.

Order 8. Eurotiales. The fruit body is a closed perithecium, that varies in size from microscopic to nearly an inch across. The asci are not regularly arranged within the fruit body, but either fill it completely or occur in loose pockets. The family Eurotiaceae includes both Aspergillus and Penicillium, important in the deterioration of a great variety of materials, and also in the production of several commercially valuable products. These genera usually are classified among the Fungi Imperfecti, because most of them produce conidia much more commonly than they do ascospores. Another family in this order, the Elaphomycetaceae, includes fungi whose fruit bodies resemble those of truffles. These fruit bodies, from half an inch to an inch in diameter, have a firm and warty or spiny covering and a dark interior with irregular pockets of asci. They are borne beneath the surface of the ground, usually in association with the roots of certain trees. A few species are world-wide in distribution, but almost nothing is known about them.

Order 9. Myriangiales. The fruit bodies resemble apothecia, but the asci are borne singly in cavities or locules scattered irregularly through a dense mass of hyphae in the upper portion of the fruit body. When the mycelium around the locules disintegrates, the asci elongate mightily and discharge their spores into the air. The order includes five families. Myriangium parasitizes scale insects on many kinds of trees throughout the world, and Elsinoe is a rather common parasite of plants.

Order 10. Dothideales. The fungi in this order bear perithecia on or within comparatively large and conspicuous black brittle masses of mycelium, called a stroma. The perithecia do not have definite walls, and the asci within them arise from relatively undifferentiated mycelium of the stroma. This lack of a definite wall lining the inner surface of the perithecia is the principal or only character that distinguishes this order from the Sphaeriales, to which they presumably are closely related. There is no general agreement among mycologists as to what families should be included in this order, whether the order itself is a "natural" one, or where it should be placed in relation to other orders of the Ascomycetes. Included in this order are such common fungi as the sooty molds, which form crusty black growths on various plants, especially in the tropics, and *Dibotryon morbosum*, the cause of black knot of wild and cultivated plums. Almost anyone who has been in the woods or brush of the northern United States is familiar with the large black crusty distortions on the branches of pin cherry. These are the result of infection of the twig by *Dibotryon morbosum*.

Order 11. Microthyriales. The order is distinguished by small, black, flat perithecia, or stromata that simulate perithecia, attached tightly to the surface of the substrate. The perithecium or stroma consists only of this flattened upper covering, the host or substrate serving as the base. Some members of the Microthyriales are fairly common in the tropics, but they are rare elsewhere, and relatively inconspicuous even where they are common. They are of minor significance from both the economic and purely scientific standpoints. It is of interest that one species, common on dead spruce needles in northern bogs, has been found commonly on dead spruce needles dredged up from preglacial bogs in northern Minnesota — leaves that were close to a million years old.

Order 12. Meliolales. The members of this group are parasites of higher plants, and, like some of their relatives in the Dothideales, they are known as "sooty molds." The mycelium is superficial and black, and small spherical perithecia are borne so abundantly on it that they form a crust on the surface of infected leaves. Most of the genera are tropical, but one, Meliola, is widely distributed throughout the tropical and subtropical regions of the world. It forms an encrusting growth on the surface of the leaves of various plants. The mycelium of some species does not penetrate into any of the leaf cells, and how it gets its food is a minor mystery. Other species send haustoria down into the epidermal cells of the leaves they inhabit.

Order *13*. Erysiphales. These fungi are obligate parasites of higher plants. The mycelium is superficial, growing only on the outside of the leaves, with haustoria penetrating into the epidermal cells. Upright chains of conidia arise from this mycelium during the spring and early summer, and perithecia are produced in late summer and fall. They comprise what are known as "powdery mildews." Many of the powdery mildews, such as those on grapes, hops, gooseberries, roses, clovers, apples, and cereal plants cause economically important diseases. The order contains only 1 family, Erysiphaceae, with 6 genera and nearly 50 species, which occur as parasites on a total of nearly 1,500 species of higher plants.

Order *14*. Hypocreales. The perithecia produced by the fungi in this order are brightly colored and of a relatively soft texture. Although only 2 families are included in the order, there are a large number of genera and species. They grow mainly as parasites on a great variety of plant material or as facultative or obligate parasites on plants, from other fungi on up to flowering plants. They are common throughout the world, and many of them cause economically important diseases of our cultivated plants. *Claviceps purpurea*, which causes ergot of rye and of some other cultivated and wild grasses, is one of the best known members of the order.

Order *15*. Sphaeriales. The fungi in this order have black, spherical perithecia, usually of brittle or so-called carbonaceous texture. They often have been called Pyrenomycetes, or *burnt wood* fungi, from the charcoal-like appearance of their perithecia or stromata. The perithecia open by a definite pore on the upper side, or by a pore at the tip of a long snout, and are borne singly or in groups on the surface of the substrate the fungus is growing on, or immersed in a dense layer of mycelium, the stroma. The order includes 12 families, about 400 genera, and over 4,000 species, approximately one third of the total number of species in the Ascomycetes. They occur literally everywhere and on all sorts of materials, both as saprophytes and facultative parasites. The family Ceratostomaceae is characterized by perithecia with very long, narrow beaks, from which the ascospores are exuded in sticky masses. Many of the fungi in this family are associated closely with insects. Some species of the genus Ceratostomella cause blue stain of wood, and the ascospores are commonly carried around by beetles and other insects. Some species of Ceratostomella cause diseases of plants, such as *Ceratostomella ulmi*, which causes Dutch elm disease, so destructive in western Europe, England, and the eastern United States. This fungus is spread almost solely by beetles.

Pleospora, in the family Sphaeriaceae, is almost omnipresent on the dung of various animals, especially the herbivores. It grows readily in culture, produces large quantities of perithecia within a few days, and these expel multitudes of ascospores. Under favorable conditions one can actually see the ascospores being discharged.

The family Xylariaceae includes a number of genera that grow mostly on wood. *Xylaria polymorpha* decays the roots of living or recently dead hardwood trees, including fruit trees. It bears prominent, erect, stromata up to several inches long, with a black outer crust in which the perithecia are formed. Xylaria is a common fungus throughout the temperate and tropical regions of the world. Hypoxylon is common on the bark of various kinds of trees, and *Hypoxylon pruinatum* causes a destructive canker of aspen trees. *Daldinia concentrica*, common on decaying wood, has brown or black hemispherical stromata up to an inch across, with innumerable perithecia just below the surface. The ascospores shot out from a single stroma in a single night are numerous enough to make a black deposit around the stroma. Stromata of this fungus taken into the laboratory have been observed to discharge successive and large crops of ascospores nightly for as long as three weeks. The fungus appears to have a built-in time sense, and discharges its spores only at night.

Order 16. Laboulbeniales. These are obligate parasites of insects. They have only a rudimentary and purely superficial mycelium. The spore-bearing structures form minute clusters or bristle-like tufts that resemble perithecia only remotely if at all, but they are considered to be perithecia. There are 6 families in the order. Few of these fungi are encountered by the average mycologist, perhaps partly because he does not know where to look for them, since one species recently has been shown to occur very abundantly upon the wings of a common and widely distributed ant. So far as is known, none of these unusual fungi cause any particular trouble or travail to the insects they parasitize.

Order 17. Hysteriales. This order is characterized by fruit bodies that are long, flat, narrow, and open by a median slit from end to end. These fruit bodies are midway between typical perithecia and typical apothecia: when they are closed they resemble a perithecium, when they are open they resemble an apothecium. Such a fruit body has been called a perithecium, an apothecium, and a hysterothecium. If one were to give a different name to every different type of fruit body produced by the Ascomycetes, there would be almost as many different types as there are genera of fungi. Recognizing that the fruit body produced by most of the Hysteriales is not a typical perithecium, no violence is done to biology in general or mycological taxonomy in particular by agreeing to call it a perithecium. Most of the Hysteriales are saprophytes or parasites on wood, bark, or leaves. Lophodermium and a few related genera are common parasites and saprophytes on the leaves of many different kinds of conifer trees. They form inconspicuous black fruit bodies up to a millimeter or two long just beneath the epidermis of the leaf. In dry weather the fruit body remains closed; but in moist weather it opens by a longitudinal slit, the asci are exposed, and they discharge their spores. A given fruit body may continue to discharge spores, during intermittent periods of moist weather, for at least several weeks.

Order 18. Phacidiales. The fruit bodies of the fungi in this order consist of a heavy black layer of mycelium, with the asci borne in irregularly branched furrows beneath it. At maturity, and in moist weather, the thick cover of mycelium splits along the top of these furrows, exposing the asci, which quickly discharge large numbers of spores. As soon as the humidity drops, the top closes, covering the asci with a very effective protective coating. Whether such fruit bodies are to be considered perithecia or apothecia is a moot question, but the particular name applied to the fruit body is of minor biological significance. Rhytisma, a common genus of this order, infects the leaves of various kinds of plants and produces black, slightly raised masses of mycelium up to a quarter of an inch across, with numerous crooked ridges, beneath which the asci are borne.

Order 19. Helotiales. There is no general agreement as to the characters that should delimit this order, or the fungi that should be included in it. The principal definitive character seems to be that the asci do not open by a definite lid, or operculum, but rather by a pore, and therefore the fungi in this group are known as Inoperculates. The presence or absence of a lid on an ascus is something that can be determined only by microscopic examination, and even then determination is not easy. In practice, familiarity with various genera and species of the order is a necessary prerequisite to placing them in the order. To found an order principally upon a single microscopic character, and a rather subtle one to boot, is not very good taxonomy; but once things like that become widely accepted, they are almost hopeless to eradicate. The order includes 4 families, most of them with cup-shaped apothecia. *Sclerotinia fructicola* is a common parasite of apples and some other fruits, on which it causes brown rot. Some species of Sclerotinia form a tight mass of mycelium, called a sclerotium, from which several stalked apothecia arise.

Order 20. Pezizales. In this group the asci open by a definite lid or operculum. The majority of the common cup fungi are in this order, as are those with stalked fruit bodies such as Morchella — the common morel — and Gyromitra — the saddle fungi. In southern South America an unusual fungus of this order, Cyttaria by name, grows parasitically upon the branches of trees and forms large fruit bodies, which formed an important source of food for the original inhabitants of Tierra del Fuego.

Order 21. Tuberales. The fungi in this order form spherical fruit bodies beneath the surface of the ground. The asci arise in a layer that lines the walls of irregular cavities. Commonly known as truffles, these fungi are gathered for food in some parts of Europe, particularly southern France and northern Italy. There are several different genera and many different species of truffles, but only a few of them have much gastronomic appeal.

Class Basidiomycetes

SUBCLASS HETEROBASIDIOMYCETES

Order 22. Tremellales. Most of the members of this order are what commonly are referred to as jelly fungi; the fruit bodies have a gelatinous texture that enables them to absorb water very quickly and produce a crop of spores quickly after a rainy spell. Most of them produce basidia that are divided into two or four cells by vertical walls. Eight families are included in the order. Some of these are widely distributed and common, like the genus Tremella, in the family Tremellaceae, and Auricularia, in the Auriculariaceae. Nearly all of the fungi in this order are saprophytic, growing mainly on decaying wood and bark. One notable exception is the family Septobasidiaceae, the members of which are parasitic on a variety of scale insects. The genus Septobasidium, in this family, already has been described in Chapter 4. Very few of the Tremellales are of any direct significance to us, although large quantities of certain jelly fungi have been used for food, particularly in the Orient.

Order 23. Uredinales. These fungi are commonly known as rusts. They include more than 100 genera and 4,600 species, all of them obligate parasites of higher plants. Most rust fungi are characterized by the production of several different kinds of spores in regular and unalterable succession, and many of them pass part of the life cycle on one kind of host plant and the rest of it on an entirely different and unrelated host plant. Many of them cause destructive diseases of cultivated plants, and therefore are likely to be of interest to anyone dealing with these plants or with their products.

Order 24. Ustilaginales. Known commonly as smuts, because

of the dark, powdery, dirty-looking spores, the Ustilaginales occur as parasites in a great variety of wild and cultivated plants. In nature most of them exist chiefly as obligate parasites, but apparently all of them can be cultured on artificial media in the laboratory. The smut fungi have in common the production of masses of chlamydospores, usually in the flowers or floral parts of their hosts, but sometimes on the stems or leaves. These chlamydospores eventually germinate to produce the basidiospores or sporidia. Sporidia of opposite sex fuse to form an infection hypha that invades the host. There are about 700 species of smut fungi, in 3 families. Ustilago, in the family Ustilaginaceae, and Tilletia, in the family Tilletiaceae, are common representatives.

SUBCLASS HOMOBASIDIOMYCETES

Order 25. Exobasidiales. This order includes only a single family, Exobasidiaceae, with 3 genera and 15 species. All of them occur as obligate parasites upon higher plants, mostly those in the heath family. They cause overgrowths of the infected tissue, so that galls, witches'-brooms, and other distortions are formed. The fungus at maturity forms a layer of basidia and associated sterile cells on the surface of the overgrowth.

Order 26. Agaricales (or Hymenomycetales). This order includes those fungi which bear their basidia in a definite layer distributed over variously arranged surfaces of fruit bodies. About 7,000 species are known, distributed through 6 families. Nearly all of the common large fleshy fungi known as mushrooms are in this order. The family Polyporaceae have the basidia distributed over the inner wall of pores, and are commonly known as pore fungi or shelf fungi. They are responsible for much of the decay of wood that occurs everywhere in the world; a few Ascomycetes grow in and decay wood, but by far the greatest amount of decay is caused by Polyporaceae. The Agaricaceae include those Agaricales whose basidia are distributed over the outside of gills. Over 4,000 species of gilled fungi have been described, most of them in the North Temperate Zone, although they are common in the tropics and in the Southern Hemisphere also. Some of these also grow in and decay wood, as most of the Polyporaceae do, but the majority of them grow in soil or decaying plant debris.

There is by no means general agreement as to what fungi should be included in the Agaricales, or how the order should be divided into Families, or how the genera should be arranged within families. For example, the genus Polyporus in the Polyporaceae, which includes most of those fungi with annual shelflike or stalked fruit bodies and with the basidia borne in pores on the under side, has been separated into more than 20 different genera by some students of the group. Also the genera within the family Agaricaceae have been arranged differently by different students of the group. Some of these systems place much more weight on making an arrangement that conforms to a "logical" plan, than on producing one that enables the student to identify the fungi in question.

Order 27. Hymenogastrales. These form a connecting link between the gill fungi and the puffballs and between the puffballs and the stinkhorns. The basidia are arranged in a hymenium in the young fruit body, but at maturity all trace of a hymenium is likely to be lost. The fruit bodies are of various shapes, most of them large enough to be seen with the naked eye. Four families are included, of which one, Secotiaceae, is represented by the common Secotium, which appears to be half way between the gill fungi and the puffballs.

Order 28. Phallales. These are the fungi known under the descriptive and accurate name of stinkhorns. The spores are borne in a sticky, ill-smelling, dark liquid on the tip of a stalk or on the inside of branched arms or other structures. Sixty-five species in 21 genera have been described. They are especially common in the tropics and in Australia, but a few species are found throughout the North Temperate Zone. All of them are saprophytes, growing in soil or in decaying roots of trees. None of them are of economic importance, but the group as a whole is interesting because of the unusual form of the fruit bodies and the complex adaptations they have evolved to insure dissemination of their spores by flies.

Order 29. Lycoperdales. These are the true puffballs, in which the basidia are borne in masses within an enclosing membrane. In the young fruit bodies, the basidia are in hymenial layers, but this arrangement is not apparent in the mature fruit bodies. The spores are dry at maturity and are suited to wind dissemination. There are many different kinds of puffballs, and one of them, *Calvatia gigantea* produces fruit bodies up to two feet in diameter. All of the true puffballs are edible, and some of them are delicious. All of them grow in soil or on decaying wood.

Order 30. Sclerodermatales. The basidia in this group never are arranged in a definite hymenium, even in the young fruit bodies. They much resemble the true puffballs, but are apt to be firmer in consistency. The fruit bodies of Scleroderma and some of its relatives in the family Sclerodermataceae are borne beneath the surface of the ground and are exposed by weathering or by the burrowing of animals or the rooting of pigs. They look much like truffles, but none of them are good to eat and some of them are poisonous. Podaxis, in the family Tolustomataceae, bears a large spherical spore chamber on the top of a stalk up to six or eight inches long, and is a rather common fungus in the drier parts of the world from the tropics to 40 degrees north and south latitude. Essentially similar fruit bodies of this fungus can be found in the Sahara, in the deserts of the southwestern United States, and in the deserts of Australia.

Order 31. Nidulariales. This order is characterized by basidia and basidiospores borne in small, enclosed chambers. In the Nidulariaceae the chambers are in a small nestlike structure, hence the name *bird's nest fungi* for this group. At maturity the small enclosed masses of spores may be scattered by raindrops. The nestlike fruit body is so constructed that a large raindrop falling into it is likely to cast out one or more of the spore-filled chambers with some force. These structures resemble small seeds, and they probably are mistaken for seeds by birds, eaten, then later deposited. At least the spores of one bird's nest fungus will not germinate until they have been exposed for some hours to a temperature approximating the body temperature of birds. Cyathus and Crucibulm are common bird's nest fungi that can be found nearly everywhere on small sticks, decaying corn cobs, and similar bits of debris. The only other family in the order, Sphaerobolaceae, bears small fruit bodies that have a single spore chamber, or gleba, which at maturity is shot away with force enough to carry it several yards. Sphaerobolus is a widespread genus in this family, growing mainly on animal dung. The mechanics of its "gun" have been investigated rather thoroughly, and it makes fascinating material for study.

Class Fungi Imperfecti

Order 32. Phyllostictales, Phomales, or Sphaeropsidales. The conidia are borne in small, enclosed fruit bodies called pycnidia. There are 4 families, separated from one another according to the structure and color of the pycnidia and the way in which the pycnidia are borne. Over 5,000 species have been described. They are found literally everywhere, both as saprophytes and facultative parasites. Many of the Ascomycetes, particularly those in the order Sphaeriales, bear their conidia in pycnidia, and so some of the fungi in the Phyllostictales are the imperfect stages of Ascomycetes. But a fungus that produces pycnidia readily and perithecia only very seldom may better be kept in the Fungi Imperfecti. This problem of whether to put a given fungus in the Ascomycetes or Basidiomycetes or Fungi Imperfecti is further complicated by the fact that several different genera of Ascomycetes may produce essentially similar asexual fruiting bodies. Thus one often cannot tell, from an examination of the imperfect fruit body, what the perfect stage of the fungus may be.

Order 33. Melanconiales. The conidia of the fungi in this order are borne in small fruit bodies, called acervuli, that are open at the top. There are all degrees of intergrades between a typical pycnidium and a typical acervulus; this adds to the difficulty of proper placement of a given fungus and to its identification by others after it has been placed in a given order. These fungi include about 1,000 species in a single family, Melanconiaceae. Like those of the preceding order, they are common saprophytes and facultative parasites throughout the world.

Order 34. Moniliales. The distinguishing character of the fungi in this order is the lack of fruit bodies. The conidia are borne on special stalks or develop from branches of the mycelium, but the conidiophores are not aggregated into a definite and distinct fruit body. There are about 4,000 species, in 640 genera and 6 families. Like those of the preceding two orders they grow both as saprophytes and as facultative parasites, and are common throughout the world.

Order 35. Mycelia Sterilia. The fungi in this artificial group produce no spores. There are 200 species, of which 100 are in the genus Sclerotium, characterized by the production of dense masses of mycelium, called sclerotia, which serve to tide the fungus over periods unfavorable for growth and which may also be disseminated. Some of the fungi in this group are known to be growth phases of certain Ascomycetes or Basidiomycetes, but some occur only as sterile mycelium. While this would seem seriously to limit their dissemination and survival, it apparently does not; for some of them, such as Rhizoctonia and Sclerotium, are almost ubiquitous in soil throughout the world, and are very successful plants indeed.

References

G. C. Ainsworth and G. R. Bisby, *A Dictionary of the Fungi*, 2d ed. Kew, Surrey, England: Imperial Mycological Institute, 1945. 410 pp.

E. A. Bessey, *Morphology and Taxonomy of Fungi*. Philadelphia: Blakiston Company, 1950. 791 pp.

F. E. Clements and C. L. Shear, *The Genera of Fungi*. New York: H. W. Wilson Company, 1931. 496 pp.

G. W. Martin, *Outline of the Fungi*. Dubuque, Iowa: Wm. C. Brown Publishing Company, 1950. 82 pp.

F. A. Wolf and F. T. Wolf, *The Fungi*, Vol. 1. New York: John Wiley & Sons, 1947. 438 pp.

Index

INDEX

Achlya, taxonomic position, 260
Acids from fungi: gluconic, 184, 197–198; citric, 195–197
Actinomyces bovis, 176–178
Actinomycoses, 175–178
Aerobiology, 38. *See also* Dissemination of fungus spores
Agaricaceae, taxonomic position, 270
Agaricales, taxonomic position, 270
Agaricus bisporiger, cultivation of, 190
Agaricus campestris, cultivation of, 190
Agriculture, U.S. Department of, research sponsored by: in climatology, 112; fungicidal wound dressing for trees, development of, 127; mushroom production, 180, 192; cheese molds, 187
Ainsworth and Bisby, 252
Air-borne spores, 35–39
Albuginaceae, taxonomic position, 260
Albugo, taxonomic position, 260
Albugo candida, taxonomic position, 261
Alga, in combination with fungus to produce lichens, 30, 45–48
Alternaria: spore formation by, 23–24; spores, shape of, illustrated, 31, Pl. 2; distance spores may be carried, 35
Alternate hosts of rust fungi: barberries, 37, 107, 113, 117; currants, gooseberries, 107, 111
Amanita, association of, with roots of forest trees, 59
Amanita muscaria, 210, 212, 213
Amanita phalloides, 209, 210, 211, 212
Amanita verna, 210, 211
Ambrosia beetles, 40, 74–78
Ants, fungi and, 40, 67–73: leaf-cutting, 69; cultivation of fungus gardens, 70–72
Aphanomyces, taxonomic position, 260
Armillaria mellea, 210
Ascoidiaceae, taxonomic position, 262
Ascolbolus, 240, 245

Ascomycetes: distinguishing characteristics, 254–255; classification of, 262–269
Aspergillus: spores, shape of, illustrated, 31; taxonomic position, 263
Aspergillus chevalieri, 223
Aspergillus flavus, 194, 217, 222, 223
Aspergillus fumigatus, parasitic in lungs of birds and man, 172
Aspergillus glaucus, 223
Aspergillus niger, used in production of citric acid, 196
Aspergillus oryzae, used in production of fungus diastase, 194–195. *See also Aspergillus flavus*
Athlete's foot, 175
Auricularia, taxonomic position, 269
Auriculariaceae, taxonomic position, 269

Barberry: alternate host of *Puccinia graminis tritici,* 37, 113; eradication of, 117
Basidiobolus: from dung of frogs, toads, snakes, 240; taxonomic position, 261
Basidiomycetes: distinguishing characteristics, 255–256; classification of, 269–273
Basidiospores of mushrooms, shape of, illustrated, 31
Beetles: ambrosia, 74–78; June, attacked by Cordyceps, 160; parasitized by Laboulbeniales, 161–162
Berkeley, mycological work of, 100
Biological warfare against insects, 159–160, 163–166
Biology, as a factor in history, 39, 112–113, 128–130
Birds: dissemination of fungi by, 28; pigeons and poultry, fungus parasite of, 171–172
Bird's nest fungus, taxonomic position, 272

Catalog

If you are interested in a list of fine Paperback
books, covering a wide range of subjects
and interests, send your name and address,
requesting your free catalog, to:

McGraw-Hill Paperbacks
330 West 42nd Street
New York, New York 10036